Historical Archaeology
in the Cortez Mining District

Mining and Society Series
Eric Nystrom, Arizona State University
Series Editor

Attempting to understand the material basis of modern culture requires an understanding of those materials in their raw state and the human effort needed to wrest them from the earth and transform them into goods. Mining thus stands at the center of important historical and contemporary questions about labor, environment, race, culture, and technology, which makes it a fruitful perspective from which to pursue meaningful inquiry at scales from local to global. *Mining and Society* examines the effects of mining on society in the broadest sense. The series covers all forms of mining in all places and times, seeing even an individualized mining history as an instantiation of global practices, markets, environments, and labor, and building from existing press strengths in mining in the American West to encompass comparative, transnational, and international topics. By not limiting its geographic scope to a single region or product, the series aims to help scholars forge connections between mining practices and individual sites, moving toward broader analyses of global mining practices and contexts in full historical and geographic perspective.

Seeing Underground: Maps, Models, and Mining Engineering in America
Eric C. Nystrom

Historical Archaeology in the Cortez Mining District: Under the Nevada Giant
Erich Obermayr and Robert W. McQueen

Historical Archaeology in the Cortez Mining District

Under the Nevada Giant

ERICH OBERMAYR

AND ROBERT W. MCQUEEN

UNIVERSITY OF NEVADA PRESS *Reno & Las Vegas*

University of Nevada Press, Reno, Nevada 89557 USA
www.unpress.nevada.edu
Cover photo courtesy of Northeastern Nevada Museum
Cover design by Louise OFarrell

LIBRARY OF CONGRESS CATALOGING-IN-PUBLICATION DATA
Names: Obermayr, Erich, 1948- author. | McQueen, Robert W., 1968- author.
Title: Historical archaeology in the Cortez Mining District : under the Nevada giant /
   Erich Obermayr and Robert W. McQueen.
Description: Reno : University of Nevada Press, 2016. | Series: Mining and Society Series |
   Includes bibliographical references and index.
Identifiers: LCCN 2016015689 (print) | LCCN 2016017274 (ebook) | ISBN 978-1-943859-22-1
   (cloth : alk. paper) | ISBN 978-0-87417-002-3 (e-book)
Subjects: LCSH: Nevada—Antiquities. | Cortez-Mill Canyon Mining District (Nev.)—
   History. | Archaeology and history—Nevada—Cortez-Mill Canyon Mining District. |
   Silver mines and mining—Nevada—Cortez-Mill Canyon Mining District. | Mines and
   mineral resources—Nevada—Cortez-Mill Canyon Mining District. | Frontier and pioneer
   life—Nevada—Cortez-Mill Canyon Mining District. | Industrial archaeology—Nevada—
   Cortez-Mill Canyon Mining District. | Social archaeology—Nevada—Cortez-Mill Canyon
   Mining District.
Classification: LCC F843.O24 2016 (print) | LCC F843 (ebook) | DDC 979.3/01—dc23
LC record available at https://lccn.loc.gov/2016015689

FIRST PRINTING

Printed in the United States of America

This book is dedicated to Estelle Bertrand Shanks

and Bill Rossi Englebright,

and the Cortez of their memories.

Contents

Illustrations

Preface

There's more ore. There's always more ore.

A geologist said this to me in 2006 while we were working in the Cortez Mining District. Me the archaeologist, and he the geologist, standing together on a canyon rim looking west toward Mount Tenabo—each of us seeing a different mountain. I was there looking for its past; he was looking for its future.

His words came after I said something to the effect that the Cortez mines closed in the 1930s because the ore played out. "There's more ore," he said. "There's always more ore; it just wasn't profitable to mine it anymore." He was right in making that distinction and I, too, should have recognized it. The fact that "there was always more ore" was the reason we found ourselves together that day in north-central Nevada. Miners and geologists had stood where we stood since the time of the Civil War, contemplating the mountain and its precious metal. Now, although times had changed, the ore within Mount Tenabo beckoned once again.

Archaeology is history as told through objects, or "artifacts." Archaeologists animate artifacts through interpretation. We aim to bring the past to the present and hopefully in the process make it relevant and enlightening to current experiences. For many people, such as the readers of this book, when you come across an artifact you inquisitively want to know its story: What is it? How old is it? Who left it here? Maybe in your mind you start building a story, projecting your own experience onto the object. Knowingly or not, you have just interpreted the artifact. However, it is important to remember that you might use an object differently than someone in the past. This is the conundrum for the archaeologist. To fully understand the artifact, we have to see it in context, and from that context we extract information about the object and its owner. For an archaeologist, context is everything. *Where* the artifact is located is as important as *what* the artifact is. Like many sciences, archaeology derives knowledge from vast amounts of experience and studies and from comparison of archaeological sites and objects. Interpreting the past is a shared experience among professionals, the public, and—most importantly and wherever possible—it also includes talking to the people who lived and worked, laughed and cried during those times that we study. Memory can be a powerful tool in the archaeologist's dig kit, though more often than not we find ourselves on sites long abandoned and any memories of them lost to time.

This book is the story of a small, north-central Nevada mining district and the everyday lives of men, women, and children that worked, lived, and—for some—died there. It is more than a standard history, which takes the reader chronologically through a beginning, middle, and end. Like a traditional history, we use old newspapers, maps, government reports, and photographs to provide a chronological framework, but the artifacts, the mountain, and the people who lived there provide the story.

By Nevada standards, the Cortez Mining District was modest and yet resilient. For a century and a half, beginning in 1863, mining in the Cortez District went through cycles of boom and bust common to mining everywhere in the Intermountain West. As late as 1928, it was Nevada's leading silver producer, although faltering production in other parts of the state helped in that achievement.[1] The district paid a living wage to hundreds of miners—including the unprecedented hiring of dozens of Chinese hardrock miners—and made one man, Simeon Wenban, a Bonanza King whose fortune rivaled anything found on the Comstock. But throughout its history the Cortez District remained relatively small and unassuming. It almost always had a post office and a school, but never a church, opera house, or local newspaper. Stage lines to Austin ran twice weekly, never daily. When needed, law enforcement came up from Austin or Eureka. There was always an operating mill in the district, but never more than one at any single time. The three communities of Mill Canyon, Shoshone Wells, and Cortez never fully succeeded in making the transition from mining camp to full-fledged town. That said, the Cortez District was worked almost continuously from its discovery in 1863 to the present day. It is one of the only mining districts in Nevada that can make such a claim.

There's more ore.

The impetus for this book was a new mining operation known as the Cortez Hills Expansion Project. It represents the next chapter in the 150-year history of mining in the Cortez District. However, before mining projects on public land can begin, they undergo a rigorous environmental review process. Most of the Cortez project is on public lands administered by the US Bureau of Land Management (BLM). As such, the project area was surveyed by archaeologists, along with a host of other environmental professionals. Federal regulations require the BLM to determine the effects of mining on significant historic and cultural resources and to offset any adverse effects to those resources. This is known as the Section 106 process, in reference to the applicable section of the National Historic Preservation Act. The archaeological project was approved by the BLM in consultation with the Nevada State Historic Preservation Office (SHPO), with the final results of the work presented in a multivolume technical report. The BLM and SHPO also required that the proponent generate a publication offering the general public an opportunity to learn about the archaeology and our discoveries at Cortez.

Breaking ground in 2008, the Cortez Hills Expansion Project was not without controversy, which we would be shortsighted not to acknowledge. Mining in general, and open pit mining in particular, has its critics. Opposition to the project came from Native Americans with deep-rooted concerns about its effect on the cultural and spiritual aspects of the landscape. Others opposed the project because of concerns about its environmental impacts. These groups raised significant social, political, and land use issues. They involve important and complex questions, deserving of a forum and discussion all their own. They are, however, beyond the scope and objectives of this book.

This story of Cortez is told by the 170,527 artifacts, give or take, found on 137 archaeological sites excavated in 2008 and 2009.[2] Each one of those artifacts was carefully mapped before being collected, washed, labeled, and analyzed. Some of them have much to tell us, while others say very little if anything at all. Some are unremarkable—the 4,080 collected tin cans come to mind—yet as tedious as their study may be, they are key to understanding daily life in the district. Many of the artifacts are similar to those seen at other sites in other mining districts, which aids greatly in interpreting them. Other artifacts are unique to Cortez or to certain demographic groups—the Western Shoshone, the Chinese, women, or children. Some are unique to specific work, like mining, woodcutting, blacksmithing, and brickmaking. However, contrary to what might be thought, archaeologists are not omniscient. Some artifacts were mysteries when we first laid eyes on them and remain so today.

The purpose of our book is simple. We want to share with you the results of all this looking, digging, and thinking. While our technical report is not available to the general public, the story we uncovered is just the opposite. It needs to be told, and to as wide an audience as possible. Our work was a continuation of previous research and investigations that showed the Cortez District had both a rich history and the archaeological resources to tell that history. The archaeology discussed in this book considerably expands on those earlier studies and, who knows, with time other archaeologists may come along and expand on this work too. Little did I know in 2006 that a decade later I would still be climbing all over that mountain. There is always more ore and, apparently, there is always more archaeology. And I hope there is always more curiosity about the past.

ROBERT MCQUEEN
Principal Investigator

NOTES

1. Nevada State Inspector of Mines 1927–1928, 10.

2. The technical report entitled *Mitigation of the Cortez Hills Expansion Project, Lander and Eureka Counties, Nevada* (Johnson and McQueen 2016) comprises four volumes and over 3,000 pages. It includes complete descriptions and interpretations of prehistoric and historic sites, artifacts, and features, how they were discovered and analyzed, and how they relate to the history of the Cortez Mining District.

Acknowledgments

This book has two authors but many hands got dirty in its making. First and foremost thanks and credit must be given to the archaeological crews that worked tirelessly on the Cortez Hills project. Crews worked for nine solid months in all of Nevada's worst seasons—especially memorable was its cold, deep winter. We are indebted to our crew chiefs and field staff—Chris Powell, Barbara Bane, André Jendresen, Jeff Michel, Paul Sanchez, Ashley Wiley, Andrea Grigg, Geoff Klemens, Melissa Murphy, Ellen Markin, and Pete Thorburn—and to the field techs—Ross Czechowicz, Tim and Frank Dann, Johnny Dickerson, Dave Duvall, Deanna Dytchkowskyj, Dave Field, Wendy Forehan, Vince Gallacci, Jenny Hildebrand, Chris Hogan, Lance Lamb, Jeff Lanham, Don Lieske, Jeanine Mahle, Chris Manning, Bethany Mathews, Angela Myerhoff, Janet Niessner, Ben Orcutt, Allison Parish, Alain Pollack, Shaun Richey, Jenna Sadd, Abigail Sanocki, Dustin Strupp, Dan Trepal, and Josh Whiting; thank you all for your hard work, dedication, and constant smiles. Additional staff helped with post-field production, laboratory and archival research, report writing, and graphics—Erika Johnson, Sarah Kunnen, Lila Lindsey, Linsie Lafayette, Natasha Nelson, JoEllen Ross-Hauer, Jennifer Sigler, and Geoff Smith. Rick Waterman and Shannon Hataway produced the final graphics in this book, and we are indebted to their fine work. Our sincerest apologies to anyone we might have overlooked.

Our thanks to Barrick Gold Corporation for supporting cultural resource management projects in the Cortez Mining District and for their enthusiasm for sharing the results of this work with the public. The staff at Barrick has been above accommodating and consistently patient. We especially thank George Fennemore and Kim Wolf for their unwavering support of our work and Bob Brock for introducing us to Estelle Bertrand Shanks and Bill Englebright.

BLM archaeologists Dr. Roberta McGonagle and Teresa Dixon were instrumental in getting this project through the administrative hurdles and, once rolling, keeping it moving forward.

Warmer and drier, but no less important, are the facilities and their staff that provided supporting information. The University of Nevada, Reno Special Collections, the Northeastern Nevada Museum, the Eureka Sentinel Museum, the Nevada Historical Society, the Nevada State Library and Archives, and the Nevada Bureau of Mines and Geology all have Cortez collections and graciously allowed us access and their use here. The W.M. Keck Earth Science & Mineral Engineering Museum has historic samples of Cortez ore, and UNR's DendroLab dated our charcoal specimens.

Dr. Priscilla Wegars of the Asian-American Comparative Collection, University of Idaho–Moscow graciously translated our artifacts having Chinese characters. Thank you to Dr. Richard Scott, UNR Department of Anthropology, for his analysis of Cletus. We are grateful to Ron James for introducing us to UNR Press, and to Eric Nystrom for his comments and encouragement of an early manuscript of this work. And we give a very special thank you to Estelle Bertrand Shanks, Bill Rossi Englebright, and Bill Magee for sharing their memories, artifacts, and photographs of Cortez with us.

Introduction

The handful of riders and pack mules were nothing more than tiny specks moving in and out among the sagebrush and juniper far below. They would have gone unnoticed, except you happened to look up and spot their slight, distant movements. The pack string made slow progress, seeming to hardly move at all, while the riders scattered out across the broad alluvial fans at the edge of the valley. Their quicker motion caught your eye. There must have been half a dozen, disappearing, then reappearing as they dropped into and then emerged from the small gullies at the base of the mountain. The riders sometimes stopped and dismounted, leading their horses, their eyes on the ground. They would pick up a rock and then break it apart with a hammer and examine the results. Some pieces went into canvas bags slung behind their saddles; others were tossed aside. At times they stopped what they were doing and looked up at the great mountain, where you watched, invisible to their searching eyes.

Nine men made up the company of prospectors who left Austin, Nevada Territory, in the spring of 1863.[1] Austin was the hub of the booming Reese River Mining District and served as the jumping-off point for the many prospectors who set out to explore north-central Nevada during the early 1860s. There were no novices in the company, no greenhorns wandering the territory and hoping to stumble onto a bonanza of gold or silver. These were experienced men, backed by a savvy group of Virginia City and California investors and led by Andrew Veatch, a highly regarded prospector who worked from his assay office in Austin.[2]

Veatch and his men had made their way north along the Simpson Park Mountains and into Grass Valley. With the Toiyabe Range on their left and the Cortez Mountains in front of them, one freestanding mountain would have gradually separated itself from the skyline. The emerging details of the mountain's landscape would have given them a daily measure of their progress. Thick stands of pinyon pine and juniper blanketed its lower slopes, with patches of scrub brush and bare rock above the treeline. But even from twenty miles away they would have made out the band of white rock, hundreds of feet thick, that rose out of the valley floor and angled skyward across the face of the mountain.

FIGURE I.1. Mount Tenabo. Summit Envirosolutions, Inc. Photo by Robert McQueen.

The white rock was a massive ledge of quartzite the prospectors eventually christened the "Nevada Giant," on the mountain they would call Tenabo (from the Shoshone word *Tinaba*, meaning "white rock water"[3]). Their excitement as they drew closer must have been hard to contain. Quartz and quartzite were signs of good things to come. In fact, gold and silver claims throughout the West were often filed as "quartz" claims, and this was a very impressive outcrop.

The Veatch party aimed for nothing less than the discovery of a major mining district, and they succeeded. The Nevada Giant initially fell short of expectations, although the mines that later probed it at great depth produced millions in silver. They staked their first major claims on the north side of the mountain, in Mill Canyon. By the autumn of 1863, the prospectors had organized the Cortez Mining District and joined with their investors to incorporate the Cortez Gold and Silver Mining Company.[4] A few months later they were mining on both the Mill Canyon and Nevada Giant sides of Mount Tenabo, while Simeon Wenban, an English-born member of the original party, began supervising construction of the mill that gave the canyon its name.[5]

The mining and milling in the Cortez District during 1864 was the beginning of a remarkable story. The district stayed in production almost continuously for eighty years, until the onset of World War II. For much of that time, from the initial discoveries until his death in 1901, it was under the personal control of Simeon Wenban. In the boom and bust world of Nevada mining only a handful of districts could claim this kind of success and longevity. If the unpredictable nature of the ore deposits was not enough, mine operators, bankers, investors, and businessmen typically fought endless legal battles over claims, mines, mills, and whatever profits they generated.

The Comstock's famous Big Bonanza had run its course by the 1880s; the Reese River and Eureka Districts in north-central Nevada were done by the 1890s; and the Tonopah and Goldfield booms that brought mining in the state back to life in the early twentieth century withered away by the 1920s. In contrast, some of the same

Cortez mines from the 1860s continued as mainstays of the district well into the 1920s and 1930s.

Significant work at Cortez came to an end in WWII, when the federal government restricted all nonstrategic precious metals mining in the country. The district remained essentially abandoned through the 1950s, a decade in which Cortez became the purview of a single, self-appointed watchman who guarded both the ghost town and the memory of what life was like generations before. In the 1960s, new exploration techniques and mining methods revolutionized mining in north-central Nevada, and miners and prospectors began re-exploring Cortez. By the early twenty-first century, the Cortez Mining District once again took its place as one of the most productive in the state.

Over time, the wood and stone buildings in the district's three small settlements—the initial camp in Mill Canyon, Shoshone Wells, and the Cortez townsite—and the cabins, dugouts, and campsites scattered across the slopes of the mountain fell away one by one. Rainfall washed the dirt mortar from rock walls, while the dry, high desert air sucked the moisture from wooden walls and roofs and shrank and warped the boards until they pulled themselves apart. Season after season of snow, wind, rain, and hot sun turned homes and businesses into heaps of rubble and bleached wood. On the mountain, the rock-lined dugouts of miners, woodcutters, and charcoal makers collapsed and filled with dirt and debris. New juniper and pinyon trees grew among the stumps of their harvested predecessors. The dark portals of neglected mines, hidden among weathered timbers and slumping walls, gave no hint of the maze of underground workings behind them, other than the streams of waste rock frozen in place on the slopes below. At the mill, the ruins of

FIGURE I.2. An adobe building at Cortez, 2006. Summit Envirosolutions, Inc. Photo by Robert McQueen.

brick walls and stone foundations still stood where once they enclosed the machinery that pulverized tons of rock, roasted and mixed it into a chemical stew, and drew
out its precious metals. Below the mill, a river of dried tailings choked and over-
flowed a gulley on its way to the valley floor.

We call abandoned, windswept, and dusty places like Cortez ghost towns, and
we find them endlessly fascinating. A ghost is something from the past, something
not here but not entirely gone either. The stories, lives, and memories that inhabit
empty, crumbling homes as well as grander mines and mills take hold of our imagination and do not let go. The lives and work of the people of Cortez and their single-
minded dedication to extracting wealth from solid rock remain among the ruins for
us to discover, explore, and study.

Archaeological exploration and research in the Cortez area began in Grass
Valley during the 1960s.[6] This work focused on the Native American inhabitants
and particularly on archaeological sites from the early contact between the newly
arrived Euro-Americans and the Western Shoshone. In the early 1980s, archaeologists took an interest in the area's mining history. They began their studies at the Shoshone Wells townsite, with the idea of linking it to the overall history of the area and
identifying ethnic groups among the residents, the contrasts in their lifestyles, and
the influence of Victorian culture on life in this remote mining district. They located,
described, and analyzed residential and archaeological features, and used the artifacts they found to identify the ethnicity of the inhabitants. Their work expanded in
subsequent years to include other settlements and the impacts the mining district
had on the area's pinyon-juniper forests.[7]

The 1990s and 2000s brought further research to the slopes of Mount Tenabo
and Grass Valley.[8] Archaeologists continued documenting mine and mill sites, along
with supporting industries like wood cutting, charcoal and lime production, transportation, and water supply.

The archaeological study culminated in 2008 with the multiyear cultural
resource investigation in conjunction with the Cortez Hills Expansion Project. The
mine project avoided Cortez and Shoshone Wells, so we did not excavate within the
townsites, although both remained important subjects for archival research. Our
archaeological study area covered large portions of the north and west slopes of
Mount Tenabo as well as the northern end of Grass Valley. We revisited and reexam-
ined previously discovered sites and investigated many new areas. Combined with
extensive historical research, data from our excavations addressed almost every
aspect of the district, how it functioned, and the lives of the people who made it
work. Our study took in exploratory mines and prospects; residential structures both
on the edges of the town of Cortez and in outlying areas on Mount Tenabo; wood
cutting and charcoal production locations; lime kilns; brickmaking sites where thousands of bricks were stacked and fired; pipelines and pump stations that brought
water not only to individual households but to equally thirsty steam engines and

FIGURE I.3. Archaeologists recording the ruins of a stone building. Summit Envirosolutions, Inc. Photo by Robert McQueen.

mills; the roads and trails that were the lifelines of the district's survival; and, everywhere, the refuse and debris from day-to-day life.

Our work answered many questions about material culture in one of the longest-active mining districts in Nevada. But archaeology also uncovers objects that, as we learn from them, offer us a unique connection to history. Whether it is a drinking cup, a can once filled with preserved fruit, or the experience of sitting in the corner of a newly excavated rock-walled dugout, archaeology links us to the world of the person who, perhaps a century ago, lived in that dugout or drank from that cup. Archaeology gives voice to those who will never be mentioned in the history books, memoirs, or newspapers or rate anything beyond a line on a census document or a death certificate filed away at the courthouse.

Archaeology can create an amazingly detailed version of day-to-day life in the past and the people who lived it. Who were they? How did they shelter themselves? What did they eat? What was their work like? What were their amusements? What did they spend their money on? But there is also a wider picture. The Cortez District was a complex, industrial-scale network of tasks and activities, all directed toward a single objective. How did that work? What were the parts? How did they interact? How did this network, created from scratch on a remote mountainside hundreds of miles from supplies, equipment, and workers, not to mention the market for its only product, manage to discover, mine, and mill millions of dollars' worth of silver?

And there is the overriding question with which our book starts and ends: What was it like to live in the Cortez Mining District? Hopefully, we present an answer that does justice both to the people living those lives and the fact that archaeology and history always discover new questions for every old one they answer.

Our first chapter looks at Cortez the place. Not the one the miners named after the Spanish conquistador, but the landscape that began forming millions of years ago, and the geological processes that endowed its deep rock formations with silver— leaving just enough in view to entice and fire the imagination of those miners.

The second chapter is about the first people, the ones who were here when the miners arrived and are still here today and the unique clues they left us about their way of life.

Chapter 3 presents a mining history of Cortez that takes us from the beginnings of the district in 1863 through the 1940s and WWII. It provides the narrative frame- work for the detailed stories we assembled from written words, faded maps, and those traces of material culture we were lucky enough to uncover during our fieldwork.

Chapter 4 examines the machines and processes that were used to extract silver from Mount Tenabo and that kept the Cortez District going through a century and a half of Nevada's evolving mining history.

Chapter 5 describes the far-flung and complex industrial system that reached into Mount Tenabo, brought out the ore, extracted the silver, and delivered it to the world.

Chapters 6 and 7 are about the people who made the Cortez District work, as seen through the lens of archaeology. Chapter 6 looks at who they were and chapter 7 delves into how they lived.

Chapter 8 focuses on life at Cortez during the 1920s when, to our good fortune, the study of material culture and the recorded memories of people who lived those times overlap.

Chapter 9 returns our story to Mill Canyon, where it began. We also follow a single archaeological discovery made near the canyon, how it began as a mystery, how the mystery was solved, and the unexpected look it gave us into "what life was like."

We conclude with chapter 10 and touch on the fact that after 150 years there is, in fact, still more ore. The Cortez District has come full circle and is once again one of Nevada's major producers of precious metal.

NOTES

1. Cortez Mining District claim book (n.d.). Bound volume includes constitution of the Cortez District, signed by the nine recorders: A. A. Veatch, James Wilson, Thos. D. McMasters, Joseph Schmadel, Simeon Wenban, Henry Farrell, Samuel McCurdy, Charles Durning, and John F. Cassell.

2. Andrew Veatch was described in the *Daily Alta California*, May 28, 1863, 1: "The mines were discovered about a year ago by a party of prospectors, known as the Veatch party, the

leading man in it being Andrew Veatch, a chemist and miner of some experience and reputation at Virginia City, where he had been engaged in the mill of the Central Company." Veatch had an office in Austin, as the *Reese River Reveille* of October 28, 1863, 1, noted in an unrelated advertisement. The subject's office was on Main Street, "one door south of Veatch's Assay Office." A canyon just outside Austin was also named after him.

The *Reese River Reveille*, October 28, 1863, 1, described the locators and their backers: "The district was discovered, located, etc. by a company gotten up in Virginia, of which John Leavitt, L. A. Booth, N. A. H. Hall, L. W. Ferris, and other old Sacramentans are members. On organization of the company, a party of ten contributed $4,000 and placed it in the hands of Andrew Veatch, a noted prospector. The conditions on which he received the money were, that he was to take eight men with him and proceed in search of a rich mining district. [With the discovery] The interests of the original company and the prospectors were united, and an additional sum of $6,000 was raised for prospecting the ledges more fully. The result was so flattering that the company was incorporated, and the machinery for a quartz mill was procured and is now being shipped by teams as above stated."

John Cassell's obituary, *San Francisco Call*, July 3, 1893, referred to him as a "well-known mining man." His projects and interests included Tuscarora, Meadow Valley Mines at Pioche, and Mono County Mines. He was credited with introducing giant powder and single-hand drilling in Idaho. Bancroft's *Hand-book Almanac for the Pacific States, 1864* (1889, 298), edited by William Henry Knight, noted that Nathaniel A. H. Ball was a delegate to the Nevada Constitutional Convention, Leonard W. Ferris was a Storey County probate judge (page 299), and John Leavitt was supervisor of the Winfield Mill and Mining Co. in Cedar Ravine, in the Comstock District (page 299).

3. Rucks (2000, 11; 2004, 37; 2008, A-32). Julian Steward's consultants identified the area near Cortez as *Tinapa* "('a white rock' + 'spring')" (Steward 1938, 142). However, the consensus among Rucks' Native American consultants was that Tenabo, rather than being a version of *Tinapa*, was an anglicized version of *Dinabo*, which means "rock writing." The white cliffs, seen as a mark or "writing," were the namesake for the mountain (Rucks 2000, 11). Contemporary Euro-American accounts and newspaper articles (e.g., *Reese River Reveille*, May 3, 1864; *Daily Reese River Reveille*, February 20, 1866) maintained that "tenabo" meant "lookout" or "lookout mountain."

4. We do not know the location of the original Cortez Gold and Silver Mining Company articles of incorporation. The first stockholders meeting was apparently held sometime prior to July 20, 1863. The *Daily Alta California*, July 26, 1863, 1, advertised an "adjourned" annual meeting of the stockholders of the Cortez Gold and Silver Mining Company, to be held at Office No. 706 Montgomery Street, San Francisco, on August 14, 1863. The notice is dated July 20, 1863. The advertisement was reprinted in the August 17, 1863, *Daily Alta California*, 2, even though the meeting date had passed.

5. *Reese River Reveille*, April 7, 1864; *Reese River Reveille*, May 7, 1864; *Reese River Reveille*, May 5, 1864: "Mr. Cassell, the general manager, Mr. Wenban, the superintendent, and Frank Doherty, the bookkeeper of the company are the right men in the right place."

6. Clewlow and Rusco (1972); Clewlow et al. (1978).

7. Hardesty and Hattori (1982, 1983, 1984); Hattori and Thompson (1987); Hattori et al. (1984).

8. This was the beginning of cultural resource management investigations associated with mining projects in the Mount Tenabo and Mill Canyon area. Some of the more substantial projects included McCabe and Reno (1994); McCabe (1996); McCabe and Obermayr (2003); McQueen et al. 2008; 2015).

Historical Archaeology
in the Cortez Mining District

Chapter 1

The Place

Archaeologists carefully map and document the sites we discover and record the precise locations of the artifacts and features we find. Recording this basic information is the first step to understanding the relationship of each artifact and feature to other artifacts and features, to their surroundings, and ultimately to history. Context also means place. In the Cortez Mining District, geography, geology, climate, and plant and animal life create the setting for life—and for historical and archaeological study. Place ruled the lives and purpose of the miners and mill workers, and the people of the district who either supported or depended upon them. Its location determined how far each tool or bite of food needed to travel to reach them. Its climate dictated how they would shelter themselves from the elements, when they could work and when they could not, and where they would get water to drink. Their success, beyond producing silver, depended on how well they took advantage of the resources Cortez offered—and how well they compensated for the ones it lacked.

The Cortez Mining District covers the north and west slopes of Mount Tenabo, at the intersection of the Cortez Mountains and the north end of the Toiyabe Range. Mount Tenabo, at 9,153 feet, is the highest and most prominent peak in the area. The district also includes the Cortez and Toiyabe foothills that form the divide between Crescent and Grass Valleys.[1]

The rugged Cortez Mountains, the Toiyabe Range, and the valleys on both sides are typical of the basin and range topography of the Great Basin. The Great Basin covers approximately 113,144 square miles of the western United States[2] and encompasses almost all of Nevada. It is a vast enclosure with no opening to the sea for its rivers. Instead, they end as sinks or terminal lakes. North–south mountain ranges and long, wide valleys typify its geography. In the Cortez area, Crescent Valley separates the Cortez Mountains and the Shoshone Range. Scattered seasonal drainages trend in the direction of the Humboldt River but disappear well short of it. To the south, Grass Valley lies between the Cortez and Simpson Park Mountains on the east and the Toiyabe Range to the west. It is an enclosed basin, with a playa—or dry lakebed—on the valley floor that absorbs the few ephemeral streams flowing down from the surrounding mountains.

The west central Great Basin is an arid region of cold winters and hot summers. The Sierra Nevada and other mountain ranges draw moisture from storms tracking their way east from the Pacific, creating a rain shadow that accounts for the basin's arid climate. Cortez receives about 10 inches of precipitation per year.[3] Snowmelt

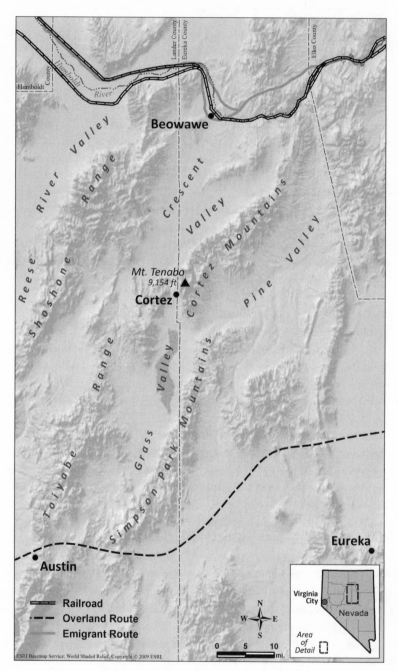

Railroad
Overland Route
Emigrant Route

N
W E
S

0   5   10
mi.

Virginia
City
Nevada

Area
of
Detail

ESRI Basemap Service: World Shaded Relief, Copyright © 2009 ESRI.

FIGURE 1.1. Geography of the Cortez region. Summit Envirosolutions, Inc. basemap courtesy of Environmental Systems Research Institute, Inc.

and a few springs supply the seasonal or rare perennial streams that flow out from the mountains and disappear in the porous, sandy sediment at the edges of the valleys. In wet years, ephemeral lakes sometimes form at low points on the valley floors, where impervious layers of silt or clay block the downward movement of the water.

These arid conditions presented challenges to both the Native American and the later Euro-American inhabitants. The Western Shoshone organized their seasonal travels to include stops at springs and streams, as the extensive prehistoric archaeological sites found at these locations show. The Euro-Americans located their mills and settlements near water whenever they could. Mountain ranges were the most likely setting for bedrock exposures of valuable minerals, and in the wider picture they were "islands of moisture" among the dry valleys.[4] But in some cases, including at Cortez, Euro-Americans had to haul water by the barrelful on the backs of mules to their most important mines and settlements. In the 1880s, it took an expensive, complicated system of pipelines, siphons, wells, and pumps to bring an adequate water supply to the new Tenabo Mill and Cortez townsite. The dry, high-altitude climate also made the Euro-Americans completely dependent on the outside world for the bulk of their food supply. The short growing season and lack of water severely limited farming and raising fruit or vegetables on a commercial scale. Local ranchers produced hay and beef for the Cortez District, but staples, canned foods, and most fresh produce had to be freighted from Salt Lake City, Sacramento, or San Francisco to either Beowawe or Austin and then hauled to Cortez.

The uncompromising climate limited the diversity of plants and animals in the district. Shadscale, greasewood, saltbush, and saltgrass covered the valley floors. Sagebrush dominated the valleys and alluvial fans, with pinyon-juniper woodland on the higher slopes and low sagebrush on the rocky upper reaches above the treeline. The Western Shoshone made the best of what the austere environment had to offer by traveling the area in a seasonal round that put them in the right place to harvest plants as they ripened and hunt a variety of animals as they followed their own seasonal migration patterns. Plants included seed-bearing grasses, pinyon nuts, and root plants such as yarrow, balsamroot, *yamba*, and bitterroot. Animals included pronghorn, deer, mountain sheep, small game, and waterfowl from the valley bottom wetlands. The Euro-Americans had no interest in native plants as a food source, aside from pinyon nuts which they often purchased from the Western Shoshone. They also hunted, but only to supplement their diet of store-bought foods.

Native Americans and Euro-Americans both valued the pinyon pine but for entirely different reasons. Pinyon nuts were a mainstay of the Western Shoshone diet, providing much of the winter food supply. They were harvested in the fall, roasted, hulled, and ground into flour. The flour was either mixed with water and eaten as a paste or soup or made into cakes for storage. Unopened pinyon cones, their nuts still inside, were also cached for use later in the winter. The pinyon harvest was an important cultural event for the Western Shoshone, complete with ceremonies, trading, and socializing that drew together different bands from all over the region.[5]

To the Euro-Americans, pinyon pine and juniper, its woodland companion, meant fuel and heat. They used juniper for cord wood to warm their homes and buildings and to fire the steam boilers that powered all their heavy machinery. They cut pinyon and converted it to charcoal to produce the high heat required for treating and smelting ore. And as we will see, it took scores of woodcutters and charcoal makers to fill this critical energy need.

The Cortez landscape started forming millions of years ago. Plate tectonics generated the geological processes that, from miles beneath the earth's surface, ultimately gave rise to today's basin and range topography. As the Pacific Plate slipped under the North American Plate, the high temperatures and increased pressure in the Earth's mantle caused its constituent rock to begin melting, although at a different rate depending on its chemical composition. Silica, for example, has a relatively low melting point, so silica-rich rock begins melting first. The hot, liquefied rock is much less dense than its surroundings. Consequently, it rises as a body and pushes upward on the Earth's crust, causing it to bulge and stretch. The stress fractures the crust and creates faults. The blocks of crust between the faults slip downward, while the adjacent areas either remain in place or continue rising. The sunken blocks, called *grabens*, become the region's valleys while the uplifted areas, or *horsts*, form the adjacent mountain ranges.

In the Cortez District, the Crescent Fault runs roughly northeast to southwest along the south edge of Crescent Valley. It marks the interface between downward-slipping Crescent Valley and the uplifting Cortez Mountains. Geologists estimate there has been a combined 10,000 feet of vertical movement along this fault, which is significant but not unusual for the Great Basin. The Cortez Fault marks the west edge of the Cortez Mountains and the east edge of Grass Valley. It is part of an extensive network of northwest-trending faults comprising the Cortez Rift. The rift reaches all the way from southern Oregon, through Cortez, to southern Nevada. It has also experienced thousands of feet of vertical displacement.[6]

The landscape we see today represents a combination of these geological processes plus thousands of years of weathering, erosion, and redeposition. The barren, uplifted mountains are shedding sediment, and gravity, rain, snowmelt, flashfloods, and the rare perennial stream carry it down through canyons and ravines to the lower elevations. The outflow forms alluvial fans that merge to create the broadly sloping plain between the base of the mountains and the valley bottom. As the gradient flattens out, the water loses its force, and by the time it reaches the valley floor it carries only the finest particles of silt and clay—the basic playa and valley floor soil types. These processes have been working on the Cortez Mountains for approximately 14

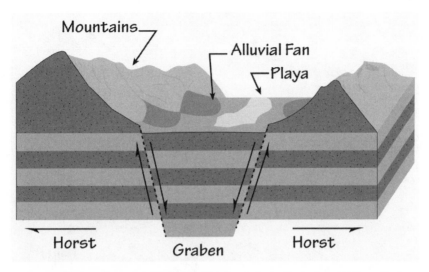

Mountains

Alluvial Fan

Playa

Horst

Graben

Horst

FIGURE 1.2. Basin and range geological and landscape formation processes. Summit Envirosolutions, Inc. Tangerine Design & Web.

million years, resulting in an accumulation of at least 10,000 feet of sediment in Crescent Valley and an equal amount in Grass Valley.[7]

The last 14 million years, though they did leave their mark on the landscape, represent a relatively recent and brief moment in the geological history of the Cortez District. The underlying geology, the rock the prospectors stared up at in 1863, is infinitely older. The Cortez Mountains are composed of a sequence of sedimentary rocks, namely dolomite, limestone, quartzite, and shale. These formed in the shallow sea that marked the western margin of North America roughly 490 to 540 million years ago, during the Cambrian Period.[8] Geological forces within the Earth's crust, at work for hundreds of millions of years, do have the power to transform seabeds into mountaintops. And along the way, molten, pressurized igneous rock is constantly on the attack, sometimes forcing its way into the smallest cracks and crevices and sometimes laying a mantle of lava rock over the land.

The Nevada Giant, easily the most prominent geological feature on the mountain, is Eureka Quartzite to geologists.[9] One early newspaper article speculated that if the Nevada Giant was a typical paying quartz vein it would contain enough silver to flood the market and cause a worldwide price collapse.[10] Years of tunneling and deep exploration did reveal ore bodies within the Eureka Quartzite, but the truly rich deposits would be discovered in the limestone adjacent to it.

The Veatch party began by developing their claims in Mill Canyon, on Mount Tenabo's north slope. Initially, only a few claims were filed in the area of the Nevada Giant. They began with the discovery of "float" or loose rock showing silver-bearing

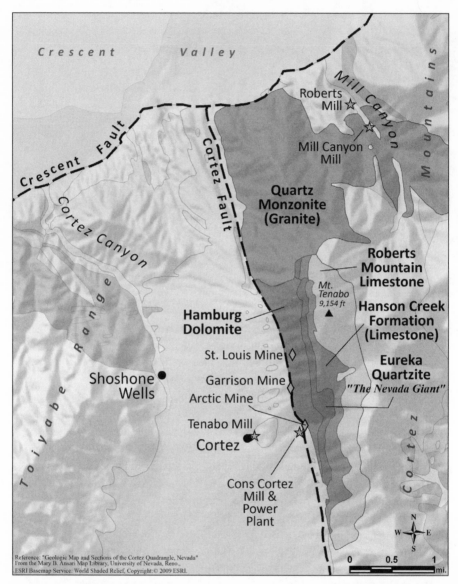

FIGURE 1.3. Geological formations, faults, and major Cortez mines. The St. Louis, Garrison, and Arctic Mines penetrated eastward into Mount Tenabo to intersect the Eureka Quartzite (Gilluly and Masursky 1965). Summit Envirosolutions, Inc. basemap courtesy of Environmental Systems Research Institute, Inc.

minerals such as quartz, pyrite, or silver chloride. Then, like following a trail of breadcrumbs, the prospectors tracked the float to its source in the side canyons or on the mountain slopes. In other cases, they were lucky enough to spot exposed veins in the bedrock itself.[11]

When the limestone, quartzite, and other sedimentary rocks that make up Mount Tenabo first formed, they did not contain precious metals. Nor did the igneous rock that later intruded into them. However, these injections of molten rock were the first step in the eventual *hydrothermal* deposition of silver and gold. Deep within the Earth's crust, pressurized molten rock searches constantly for avenues of release in the overlying formations. It seeks out weak points, sometimes making its own fractures and sometimes working its way between the bedding planes or layers of sedimentary rock. When these intrusions cut through the bedding planes they create *dikes*, which are essentially large veins of solidified igneous rock. Super-heated water is also trapped and pressurized along with the molten rock. It exploits the breaks in the sedimentary structure and forces its way into the various fractures along the dikes or faults. If it is acidic enough, it can even carve out chambers by dissolving the calcium in sedimentary rocks like limestone. Pressurized, super-heated water can also dissolve silver or gold-bearing minerals, which it carries into the fissures and chambers. Sooner or later the superheated water starts to cool, and as it does the dissolved minerals begin precipitating. Precipitation can also occur as a chemical reaction, in which the dissolved mineral interacts with elements in the surrounding rock and solidifies. Over many thousands of years, and under just the right temperatures, pressure, and geological conditions, the precipitated minerals transform cracks in the rock into veins of precious metal or the larger cavities into lode deposits.

On the north side of Mount Tenabo, a massive granitic intrusion cut through the bedding planes in the limestone. It sent out small "tongues" into the surrounding rock, which superheated water then transformed into veins and irregular shoots of silver-bearing minerals. As Mill Creek then cut its way through the limestone and granite it also exposed the silver ore, laying it open for the inquiring prospectors and their picks and shovels. The largest ore bodies on the Mill Canyon side of Mount Tenabo were sometimes 20 feet wide and hundreds of feet long.[12]

On the Nevada Giant side, decades of exploration and mining deep within Mount Tenabo found silver ore in irregular-shaped chambers within the limestone or in long, ribbonlike deposits between the bedding planes. Silver also precipitated in what miners called "chimneys"—in dikes and the fissures that formed along with them.

NOTES

1. Boundaries for the Cortez Mining District were never established. Article II of the 1863 district bylaws states: "We believe other Districts will be formed adjoining the Cortez District;

therefore, we will describe no limit for the Cortez District, but will in the event of other Districts being formed, hold claim to ten miles square as the Cortez District, the crest of Bullion Hill, at the Andrew Veatch Ledge being the center of said District." Tingley (1992) presents a generally acknowledged boundary for the district.

2. "Great Basin Information," accessed February 15, 2016, http://greatbasin.wr.usgs.gov/.

3. Desert Research Institute (2013), Monthly Precipitation, Cortez Gold Mine, Nevada; Natural Resources Conservation Service, US Department of Agriculture (n.d.), *Nevada Annual Precipitation.*

4. Wycoff and Dilsaver (1995, 139).

5. Rucks (2004, 11–12).

6. Bureau of Land Management (2008, sections 3.1–5).

7. Bureau of Land Management (2008, sections 3.1–2).

8. Bureau of Land Management (2008, sections 3.1–1 through 3.1–5); Roberts et al. (1967, 54–55).

9. Gilluly and Masursky (1965a, plate 1), *Geologic Map and Sections of the Cortez Quadrangle, Nevada.*

10. *Reese River Reveille*, May 3, 1864, 1.

11. *Reese River Reveille*, May 5, 1864, 1: "A large number of fragments of rock, containing sulphurets and native silver, were seen on the hill-sides and ravines, which being followed up led to the discovery of the St. Louis."

12. Hezzlewood (1930, 2–3); Vanderburg (1938); Emmons (1910).

Chapter 2

One Story

According to archaeologists, human beings first saw the mountain slopes, alluvial fans, canyons, and valleys that became the Cortez Mining District more than 10,000 years ago. The prehistoric archaeological account begins then, and it goes on to explain how during the ensuing millennia the world changed and people changed with it.[1] Archaeological study begins with simple curiosity about fascinating objects: arrowheads, baskets, pottery, and, yes, old cans and bottles. We find them today, in the present, but they are from time past. For *prehistoric* archaeologists, this past begins with the ancestors of *Homo sapiens* and only ends when people started putting their history in writing. *Historic* archaeologists in the Great Basin generally begin their study with the first contact between Euro-Americans and Native Americans. Our book follows the written and historic archaeological versions of the Cortez Mining District, but in the course of our work we also encountered a number of prehistoric archaeological sites. Though small in number, these sites deepened our understanding of the district's setting. Andrew Veatch, Simeon Wenban, and the other prospectors were not the first people here, nor were the miners, mill workers, merchants, or families who followed ever only people. They were newcomers, and our story would be incomplete without including what we know of the world they entered into—and changed forever.

In Europe, curiosity about objects from the past sent the first archaeologists to Egypt, the Middle East, Greece, and Rome to uncover awe-inspiring monuments and architecture and to fill museums with works of ancient art. Transposed in the twentieth century to the Great Basin, this same curiosity fueled the search for much less spectacular finds, but with a new, scientific approach. In 1924, archaeologists investigating Lovelock Cave discovered the ancient tule duck decoys preserved there. Excavations in the 1930s at Gypsum Cave near Las Vegas and Tule Springs, also in southern Nevada, found stone tools thought to be associated with the bones of giant ground sloths and other extinct animals. Later work, benefitting from decades of progress in archaeological science, set aside these initial interpretations, but at mid-century, archaeologists found human remains in Spirit Cave, Nevada, dating to 9,000 years ago. These are still counted among the oldest ever found in North America. In the 1970s and 1980s, archaeologists working in Monitor Valley used radio carbon dates to show how changing styles of projectile points could serve as markers for different time periods. Building on this research, archaeologists have pieced together an account of Great Basin prehistory that grows richer and more comprehensive with every new project. They have spent careers, and lifetimes, creating hypotheses and theories, testing them, rejecting some and accepting others, and then using the results to ask more questions and formulate new ideas. Their methods are rooted in Euro-American culture, which, like cultures everywhere, guides us in explaining where we, as human beings, came from and how we got here. But there are different cultures at work in the Great Basin. They have different answers to those questions, and in fact different reasons for asking them in the first place.

Long ago, when all the birds and animals still walked about and talked like humans, two women—a mother and daughter—gave Coyote a large basketry water jug, sealed with pitch and with a cedar knot for a stopper. They told him to take it to the center of the country, but gave him specific instructions not to open the jug until he got there. Coyote started his journey but soon got tired of carrying the jug and became curious about what was in it. So he stopped occasionally and opened the stopper to look inside. The jug contained all the people, and some escaped each time he opened it and they became the various groups of Shoshone who live throughout the Great Basin. Or did the Creator shape the first man out of dust, breathe life into him, and then make the first woman from one of his ribs? Have the Shoshone been here since time immemorial—since a time beyond the reach of human memory? Or can human beings' time in the Great Basin be measured by the rate of decay in carbon atoms and divided into stages of an "archaic" way of life? And what, exactly, is the difference between time immemorial and 10,000 years?[2]

THE PRE-ARCHAIC

Regardless of whom we think they were, the first people to see Mount Tenabo explored a world completely unlike the high desert of today. The environment was colder and wetter. Sagebrush steppe vegetation blanketed the lower elevations, while a few pines marked the uplands. But the most startling difference during this time, which archaeologists have designated the Pre-Archaic, was the lake covering the floor of Grass Valley that stretched from the foot of Mount Tenabo south as far as the eye could see. Unrelenting rain, snow, and cold had left the valley bottoms of the Great Basin submerged under fresh water lakes or covered by vast marshes and wetlands. The valley beneath Mount Tenabo was no exception.

These first people were highly mobile hunters and foragers. They traveled in small groups, making camp around the lakes that figured so prominently in their environment. There is little evidence of their passing to catch the eye of archaeologists. We have rare examples of the fluted spearpoints used by the some of these early inhabitants of North America, but only one is known from this part of Nevada. It was found in Crescent Valley.

Later in the Pre-Archaic archaeological record, more numerous types of large stemmed and concave base spearpoints replace the fluted points. They are often found on the margins of extinct lakes and marshes, but they also mark a time when people increased their range and pursued a broader variety of plant and animal foods. The lakeshore and wetland areas and the steppe vegetation surrounding them were rich with water fowl, fish, game, and various edible plants. But later Pre-Archaic sites are also found in caves, on stream terraces, and in mountain meadows and other upland areas. In many of these locations, the accumulating soil preserved plant and animal remains rarely found at the more exposed sites. These were typically jack rabbit, deer, mountain sheep, pronghorn, and occasionally bison bones. The Pre-Archaic diet also included various seeds, like Indian ricegrass and sunflowers, and freshwater fish and mollusks at sites near water.

Evidence of Pre-Archaic habitation of the Cortez area comes from two sites: the West Sinter Quarry Site approximately half a mile northeast of Mount Tenabo, and the Knudtsen Site in Grass Valley. We added our own examples with the discovery of stemmed and concave base Pre-Archaic points at four sites closer to Cortez. Three were on the old Lake Gilbert shoreline, and one was at the foot of a nearby alluvial fan. One lakeshore site was buried two feet under the surface and included a number of other flaked stone tools and large mammal bones.

EARLY ARCHAIC

Archaeologists have divided the Archaic—that is, the prehistoric Great Basin after about 8,000 years ago—into three parts: Early, Middle, and Late. These divisions are not as simplistic and arbitrary as they first appear. They represent significant changes in both the environment and human populations and the consequent impact these changes had on the way people lived.

The generous Pre-Archaic environment turned stingy 8,000 to 7,500 years ago. A suddenly warmer, drier climate brought on a drought that gripped the Great Basin for the next 4,000 years. The lowland lakes and marshes dried up, salt brush took over the sagebrush steppe, and scarcity replaced the bountiful food resources the wetter climate supported. Archaeologists theorize that the stress of the drought not only reduced the population but forced people to break up into smaller and smaller groups, search more widely across the landscape, and settle for relatively "low value" foods. One example was the reliance on various grass seeds, as shown by the numerous grinding stones in sites from this period. Gathering, grinding, and preparing grass seeds took more time and effort than root collecting or hunting, but the drought had taken a toll on these "high value" resources, leaving people with little alternative but to work harder for less of a return.

People abandoned large areas of the central Great Basin at this time or kept to a few refuges near surviving springs or streams. Six Early Archaic components were found during our excavations. A component consists of artifacts representing a certain time period. A site can include more than one component if it was repeatedly occupied over a long period of time. It could be that the Cortez area was one of these refuges, but six components over a span of 4,000 years come close to representing a land empty of human beings.

MIDDLE ARCHAIC

The drought broke about 4,500 years ago, marking for archaeologists the transition from the Early to the Middle Archaic. A cooler, moister climate returned, although not to the extremes seen during the Pre-Archaic and earlier. Lakes and marshes again appeared on the valley floors, coniferous forests expanded down from the higher elevations, and populations of large ungulates (deer, elk, and mountain sheep) increased. Pinyon nuts became a staple for many inhabitants of the area south of the Humboldt River. Pinyon pine had been expanding its range northward and reached the central Great Basin about 6,500 years ago, but dense pinyon woodlands probably did not evolve until conditions improved during the Middle Archaic.

The changed environment meant lusher vegetation, more animals, and ultimately more people. The archaeological record shows a tremendous increase in the number of sites from this period, as the growing population formed large, semipermanent villages along rivers and other lowland areas. They also began utilizing higher-altitude areas, hunting and collecting roots and other plant foods from upland base camps. Material culture, house construction, and ceremonies became more elaborate. Marine shell ornaments even reached the Great Basin by trade from the Pacific Coast. It was a time when people maximized use of the full range of resources their environment had to offer and established the Native lifeway that existed at the time of Euro-American contact.

Research has shown that the inhabitants of the Cortez region likewise established lowland villages, upland root camps, and hunting sites. The James Creek Phase, from about 3,500 to 1,400 years ago, shows the continued growth in population density. Our work identified fourteen substantial James Creek components, marked by ubiquitous Elko projectile points and numerous grinding stones.

LATE ARCHAIC

The Late Archaic spans the time from about 1,300 years ago to the arrival of the Euro-Americans. It included harsh periods of drought, but these were short in comparison to the widespread Early Archaic drought. The population continued to increase through the Late Archaic, with some interruptions probably marking temporary returns to drier climatic conditions. The numerous archaeological sites from this period reflect the growing population. Archaeologists have divided the Late Archaic into two phases in the central Great Basin, representing variations in the overall cultural pattern. The Maggie Creek Phase lasted from 1,400 to about 600 years ago, with the appearance of small, chipped stone arrow points in the archaeological record marking the new use of the bow and arrow. It was also apparently a less hospitable time in the Cortez study area, as we found only four components dating to this phase. (Remember, however, the Late Archaic phases include relatively short time periods. Four components over an 800-year period does not represent complete abandonment of the area, in contrast to the six components we found representing the 4,000-year span of the entire Early Archaic.) The Eagle Rock Phase went from 600 or so years ago to the first signs of the approaching Euro-Americans. At Cortez, as in the surrounding region, the numerous, complex archaeological sites from this phase demonstrate a continuing growth in population. Pottery also made its appearance in the Eagle Rock Phase.

The bow and arrow appeared in the central Great Basin between 1,800 and 1,300 years ago, and pottery made its appearance about 600 years ago. Both marked an intensified use of a widening range of resources, with pottery seen as signifying a turn toward a less mobile lifestyle.

The bow and arrow improved and diversified the quest for game. Hunters could shoot arrows farther and more accurately than they could throw a spear, even with the aid of an atlatl. They could also rapidly fire off numerous arrows from a concealed position, without giving themselves away. This not only improved their chances against deer and large animals, it also gave them a new and effective way of taking rabbits, squirrels, and other small mammals and birds.

Great Basin pottery was made by building up a vessel with coils of clay, smoothing them over, and then burnishing the exterior. They were fired at a low temperature, but became durable, watertight containers. Studies have shown that pottery could be used for boiling seeds by setting the pot directly onto a fire, rather than using the hot rock method, in which heated rocks were dropped into a water-filled basket to bring the liquid to a boil. Using pots to prepare meat stews also retained nutrients that were lost cooking meat over an open fire.

PAST AND PRESENT

A century of archaeological research in the central Great Basin has built a "prehistory" for the region, drawing upon data from archaeological sites, artifacts, and features, and their place in the landscape. It is based on understanding the part each site played in a wider system and, more importantly, recognizing and understanding what the changes through time in those sites mean. Clearly, people were in the Cortez area as early as anywhere else in the region. And archaeologists have recognized at Cortez and across the Great Basin that as the drought broke 4,000 years ago people developed the seasonal round that was in place when the first Euro-Americans arrived. The Great Basin has a rich variety of plant and animal resources,

FIGURE 2.1. Projectile points from our excavations include Pre-Archaic stemmed and concave-base points (*top row*); large side-notched points from the Early Archaic (*middle row, left*); a Rose Gate point (*middle row, center*) and a typical Elko Point (*middle row, right*); and arrow points from the Late Archaic (*bottom row*). Courtesy of Summit Envirosolutions, Inc.

FIGURE 2.2. Mano (*left*) and pestles (*right*). Manos were used to grind grass seeds on the flat surface of a metate; pestles were used with a mortar for grinding and mixing. Courtesy of Summit Envirosolutions, Inc.

but they are also scattered across the landscape and, individually, only ripe or ready for consumption at specific times of the year. Pinyon pine, for example, is restricted to a particular zone in the middle elevations of mountain ranges. Pine nuts can be collected green in late summer, but the most productive harvest is limited to a few weeks in the fall. Game animals travel constantly across the land, sometimes crowded together in herds along predictable migration routes and at other times nowhere to be found.

From the beginning, Native people have read and deciphered the clues that told them where to be and when to be there. Even as the climate changed, sometimes radically over long periods of time, people adjusted. They took every opportunity their culture and technology gave them to make the most of their environment. This was the Western Shoshone way of life at Cortez since "time immemorial." Families wintered near their food caches in the shelter of river valleys and lowland areas along the Humboldt River near Beowawe, and in Grass Valley. They also occasionally spent winters in *Tinaba's* well-known pinyon woodlands. They dispersed in spring to gather various kinds of newly sprouting greens and capture woodchucks, who were still fat from hibernating, squirrels, and sage hens. They dug roots, drying

FIGURE 2.3. An Eagle Rock Phase hearth, consisting of a loose ring of fifteen fire-broken rocks. The surrounding ashy soil contained burnt bone fragments and charred sagebrush, fragments of large mammal bones, goosefoot seeds (an important food resource), and a Desert Side-notched arrow point. The hearth was radio-carbon dated to about CE 1660, when Euro-Americans were well established in North America and Old Mexico but unknown to the Western Shoshone. Courtesy of Summit Envirosolutions, Inc.

and storing them in pits, caves, and rock crevices, and often walling them in with rocks, brush, and mud. The cliffs on the west face of *Tinaba* were said to be the site of many such caches.

Later in the summer, people established base camps in well-watered locations like Cortez Canyon, or Horse Canyon in Mount Tenabo's southeastern foothills. Men hunted and fished the mountain streams, while women gathered seeds, roots, tubers, and bulbs. In late summer, buffalo berries, currants, service berries, and choke-cherries ripened. Pinyon nuts were harvested in the fall and then prepared and consumed or stored for winter. Fall was also the time for communal rabbit drives and more hunting once the pinyon harvest was completed.[3]

Then everything changed. First came trade goods, like beads and iron tools, passing through many hands and reaching the central Great Basin by the late 1700s and early 1800s. Unexplained diseases appeared too, transmitted along with stories of strange new people. At first, the Euro-Americans came in small groups, trapping and trading for beaver pelts. Then came the emigrants and, finally, the prospectors and miners who touched even the remotest corners of the Western Shoshone world in their tireless, obsessive search for shiny metal.

NOTES

1. The prehistoric archaeological background is synthesized from a number of sources. Elston (1986) in his "Prehistory of the Western Area" in *Handbook of North American Indians, vol. 11, Great Basin*, presents the adaptive strategy approach and general time scale that still underpin Great Basin archaeological research. Descriptions of early work are drawn from Don Fowler's "History of Research," in the same volume. Grayson (2011) offers a comprehensive and updated look at the "natural prehistory" of the Great Basin and includes human history in Part 5, "Great Basin Archaeology." The most recent treatment focused specifically on the Cortez area is found in Johnson and McQueen (2016, Volume II: Prehistoric Resources). This is also the source of our site descriptions, artifact photographs, and interpretations.

2. This composite is drawn from a number of Shoshone origin stories recorded and published by Julian H. Steward (1943) and a similar version in Crum (1994, 1).

3. Johnson and McQueen (2016, Volume I: Project Overview, Research Design and Methods, Ethnographic Background); Rucks (2004).

Chapter 3

A Mining History of the Cortez District

The prospectors who signed their names to the bylaws of the Cortez Mining District in May 1863, and then spent the summer scrambling over the slopes of Mount Tenabo and the steep walls of Mill Canyon, began one of Nevada's most exceptional mining stories. The district's history started with the first mines and mill in Mill Canyon in 1863, and it essentially ended there eighty years later, at the beginning of World War II, with the closing of the Roberts Mill. During this time, the Cortez District produced $12,500,000 worth of silver.[1] The greatest production came not from Mill Canyon but from the Nevada Giant side of the mountain. It turned one man into a millionaire, gave work to many more, and was a place where countless others lived their lives and made homes and families.

The Cortez District has an important place in the study of mining history. First, because of its longevity, the district spanned eight decades of progress in mining and milling technology, and because it was in almost continuous production practically every major development in the industry was put into practice at Cortez. Wenban and his successors' experiments and improvements included the Washoe and Reese River Processes in Mill Canyon, advanced lixiviation at the original Tenabo Mill, cyanide leaching and tailings reprocessing at a refurbished Tenabo Mill, and state of the art cyanide and flotation processing at the Consolidated Cortez Mill and later at the Roberts Mill in Mill Canyon.

The Cortez District also employed advances and inventions from the nonmining world that affected the way silver was mined, extracted from ore, and delivered to market. Gasoline and diesel engines replaced steam and animal power; compressed air rock drills replaced single and double jack hand drills; electric lights replaced candles for illuminating underground workings; railroads and highways replaced wagon roads; and new forms of communication and transportation made it increasingly easy to link even the most isolated mining district to worldwide finance, markets, and suppliers.

Second, the Cortez District was essentially controlled by one man, Simeon Wenban, for the better part of its existence. Wenban was a notable figure in Nevada's mining history who, although acknowledged as a "Bonanza King" in his own time, has subsequently been overlooked by history. His notoriety does not approach the legacy of the Comstock's Big Four of Mackay, Fair, Flood, and O'Brien, even though his tenure at Cortez eclipsed the Comstock's initial success in the early 1860s, the Big Bonanza of the 1870s, and the ultimate bust of the 1880s. Wenban not only outlasted the Comstock but spent the fifteen years from 1886 to 1901, with the Comstock mired in depression, leading Cortez to unprecedented success. Wenban began at Cortez with the initial discoveries and organization of the mining district, and he stayed committed to its development throughout his life. He became the major owner during the difficult years of the late 1860s and early 1870s, when his fellow locators moved on and other investors either abandoned their claims or sold out. He remained the major figure for almost four decades, and he was still at work in the district when he died in 1901.

Wenban's position and longevity meant that, in contrast to other mining districts, important decisions at Cortez were made by one man. Wenban's success, and the success that continued after his death, spoke volumes for his foresight and knack for making the right choices. He recognized the potential of the Nevada Giant side of Mount Tenabo and shifted his efforts away from Mill Canyon very early in the district's history. His commitment to employing the latest, state-of-the-art equipment and processing methods in the Tenabo Mill, along with investing in a new water supply system, laid the foundation for decades of profitability.

Wenban's decision to employ Chinese, including hiring them as underground miners, was another example of his willingness to take unorthodox and even unpopular steps if they benefited his company. His long-term use of Chinese workers defied common practice throughout the West, which typically kept the Chinese out of the higher-skilled—and higher-paid—jobs.[2] It lead to the creation of a small but thriving Chinese community at Cortez and was critical in its own way to Wenban's success. The Chinese willingly worked for lower wages than any other ethnic group.

The Cortez District began as a very typical example of a story playing out all across Nevada during the 1860s. Like those other mining districts, it really began fifteen years earlier, in the late 1840s. Emigrants traveling the Carson River Route of the California Trail found gold in the sand and gravel at the foot of Mount Davidson, near the boundary between California and what was then the Utah Territory. Their discovery eventually led to the world famous Comstock Lode, but not until prospectors and miners spent the next decade exploring Mount Davidson and digging, panning, and sluicing in the canyons on its eastern slope. The nuggets and flecks of gold they found had to come from somewhere on the mountain, and for years they searched for this ultimate source, or "mother lode." They never found a mother lode of gold, but they did find silver. One of the world's richest mining districts began as troublesome blue clay that clung to picks and shovels and generally clogged up the equipment. In 1859, someone finally assayed a sample of the clay, and to everyone's surprise it tested out at $3,000 per ton of silver and $900 per ton of gold.[3] The rush to Washoe was on.

The Washoe District, more famously known as the Comstock, attracted thousands of hopeful prospectors. And it soon became the jumping-off point for others bent on exploring the jagged mountain ranges lined up one after another between Mount Davidson and central Utah. In the early 1860s, legions of prospectors spread

FIGURE 3.1. Simeon Wenban. Courtesy of Nevada
Historical Society.

eastward from the Comstock. Each man hoped to make the next big strike, and a
few actually succeeded.

A number of Nevada's most important mining districts were discovered at this
time. These included Tuscarora in the northern part of the state, Aurora in the south,
Manhattan and the White Pine District in central and eastern Nevada, and the Reese
River District, 75 miles east of Virginia City. The Reese River mines and mills gave
rise to the town of Austin and, not long after, helped open the way for Cortez to join
this illustrious company.

There were plenty of naïve, ill-equipped, and unprepared loners and spur-of-
the-moment partnerships among the gold and silver seekers. After all, riches were
there for the taking, and finding them was just a matter of wandering long enough
over the landscape. Others were much better prepared. Andrew Veatch and the men
who set out with him from Austin in the spring of 1863 had the skills and experience
to make a significant discovery, and they were not going to settle for anything less.
Veatch's backers included investors from Virginia City as well as Sacramento veterans
of the California Gold Rush. They provided Veatch with $4,000—a substantial sum
in 1863 (roughly $72,000 in 2010 dollars)—to assemble and equip his team.[4] They
also had the money and connections to follow up on any discoveries and develop
them into money-making mining enterprises. Veatch was still a young man, but he
had his own assay office in Austin and had explored for minerals in Mexico, California, Arizona, and the Washoe District. According to the *Reese River Reveille*, he was

"an excellent judge of minerals and the most expert prospector in the world."[5] The group included Veatch, Simeon Wenban, and seven other hand-picked men, who would share any future success equally with the investors.

The Veatch party struck out northward from Austin, working their way along the Simpson Park Mountains into Grass Valley. They reached Mount Tenabo in May 1863. The Nevada Giant, visible on the west slope of the mountain from miles away, was by all accounts an obvious target. It is unlikely Veatch and his men were the first Euro-Americans to see Mount Tenabo, however the early explorers either missed or did not take note of this prominent landform. Captain Simpson, who surveyed the Central Route in 1859, passed well to the south.[6] Edward Beckwith, who surveyed the route of the transcontinental railroad, crossed a pass south of Mount Tenabo without noting anything special about the mountain to his right.[7]

The Veatch party staked a number of claims on the west side of Mount Tenabo, but they had the impression that the Nevada Giant was relatively barren of precious metal. Shifting to the north side of the mountain, they located several very rich veins of silver in the granite of Mill Canyon. The move to the canyon, at the expense of the limestone on the Nevada Giant side, was based on Veatch's belief that silver was not to be found in limestone.[8] This was later shown to be wrong, but the initial discoveries in Mill Canyon looked rich and easily accessible. The silver was in well-defined veins and fissures, anywhere from 6 inches to 6 feet wide and spaced from 30 to 300 feet apart. Some cropped out conspicuously on the ground surface, but most were barely exposed to view.[9] They appeared to be much like the Comstock deposits, which would have been reassuringly familiar to some members of the party, including Simeon Wenban.

The group staked fifty-six claims and organized the Cortez Mining District at Veatch Camp on May 15, 1863.[10] They elected officers and set out rules and procedures for filing and maintaining claims. Since they had no idea of the real extent of the silver deposits, they arbitrarily described the district as 10 by 10 miles square, centered on the Veatch Ledge at the crest of what they called "Bullion Hill."

Things moved fast after this initial work, and the prospectors and their backers incorporated the Cortez Gold and Silver Mining Company in San Francisco.[11] Stockholders were a mix of new and previous investors, including some prominent members of the Virginia City community. John Leavitt, for example, was one of the first state assemblymen in the new state government that would be organized in 1864. Nathaniel A. H. Ball served as a member of the state constitutional conventions of 1863 and 1864. Leonard W. Ferris was an attorney who also eventually served as a probate judge in Carson and Storey Counties.[12] George Hearst, of San Francisco, was the best known of the early backers.[13] Washoe's Ophir Mine had provided the foundation for his considerable fortune, and while he suffered setbacks in the coming years he eventually became one of the richest men in the country. His son, William Randolph, would put together the newspaper empire that bore the family name.

In the summer of 1863, Simeon Wenban reported that the company was working four "leads," or veins, each at a depth of 16 feet. The veins were easy to follow
and showed enormous value, averaging $400 to $500 in silver per ton. The richest
ore assayed at an astonishing $3,000 per ton. They had graded a mile of road to the
mines, erected a blacksmith's forge, and laid foundations for two buildings. Fourteen
workers were employed at the site, and two families were in residence.[14]

The Cortez Gold and Silver Mining Company held its first stockholders meeting on August 14, 1863, at the company offices in San Francisco.[15] Ten days later the
board of trustees leveled an assessment of $5 per share, totaling $50,000, to fund construction of a mill.[16] Mining companies commonly used assessments to raise money
for specific projects or operating expenses if they ran short of cash. Each shareholder
was expected to contribute the assessed, per share amount, and those who did not
would potentially forfeit their stock. The mill would eliminate the need to transport
ore by mule train to the Austin mills for processing.

The Mill Canyon mill was intended to be the most complete mill east of Virginia City. (The mill at Mill Canyon was actually called the Tenabo Mill; to avoid
confusion with the later and more popularly known Tenabo Mill on the other side
of the mountain, we will refer to it here simply as the Mill Canyon mill.) The Cortez
Gold and Silver Mining Company acquired milling machinery in San Francisco and
freighted it across the Sierra Nevada to Carson City and out the Reese River Road to
Austin. From Austin, teams took the machinery north along primitive wagon roads
the length of Grass Valley, out through Cortez Canyon into Crescent Valley, and then
up Mill Canyon to the mill site. In December, the *Reese River Reveille* noted that the
Cortez Company had contracted to pay the "snug little sum" of $24,500 in freight
charges to haul 60 tons of milling machinery and equipment from Sacramento to
Cortez.[17] The components included a boiler, steam engines, stamps, steel vats and
mixers, and massive pine beams for the mill's superstructure.

A mild winter aided the effort, and the last of the machinery arrived in Mill
Canyon in April 1864. In May, the *Reveille* described the scene in Mill Canyon, where
"the machinery is now all on the ground and a large corps of mechanics and laboring men are now engaged in preparing the ground and placing the machinery in
position."[18]

The Cortez Gold and Silver Mining Company named Wenban its superintendent. He was born in England in 1824 and immigrated as a child with his family
to a farm near Cleveland, Ohio. He married in 1847 and had two daughters, Flora
and Eva. In 1854, he set out for the goldfields of California, sailing from New York
City and arriving in San Francisco in March. By this time, the gold rush of 1849 had
evolved away from independent placer mining to more industrialized hydraulic and
underground mining. Wenban spent the next six years in Sierra County, in the Sierra
Nevada, eventually turning his attention to the study of mineralogy and mining technology. The 1860 US Census listed him as a machinist, residing near Forest City,

California. By 1862, he had made his way to Washoe where he supervised a quartz mill belonging to C. B. and C. Land. Andrew Veatch apparently recruited him at this time, as Wenban was known as "a practical engineer who knew all the requirements of working a mill."[19]

In Mill Canyon, a building 95 by 50 feet housed the mill. The complex also included an ore house, blacksmith and machine shop, and water tanks. A 40 horsepower steam engine ran the machinery, with steam generated in a 15 foot long by 4 foot diameter boiler. Mill Creek provided water for the boiler and other operations within the mill. The Vulcan Works in San Francisco manufactured the engine, and Coffee & Risdon, also of San Francisco, made the boiler. The engine was set on a bed of neatly cut granitic blocks, capped with heavy timbers. It turned a massive belt wheel which then turned a main shaft. Stamp batteries, mixers, and amalgamators all ran from the main shaft, linked to it by rubber belting.[20]

The mill was up and running in mid-1864, but it immediately ran into problems. Working with the Land Brothers on the Comstock, Wenban employed a variation of a beneficiation process known as the Washoe pan process. Beneficiation is the chemical and mechanical treatment of ore to extract the precious metal from the gangue, or worthless parent rock. The Washoe process worked well, and almost every advanced mill on the Comstock used it or something similar. However, Cortez ore was not Comstock ore. The Cortez silver was "refractory" ore, meaning the precious metal was locked up in chemical compounds within another host mineral. In nineteenth-century terminology, it was "rebellious" ore. In addition, lead and other base elements occurred in much higher proportions at Cortez. The complex chemical bonds between the silver compounds and base metals resisted attempts at breaking them down. And as promising as it looked, the Cortez ore was lower grade than Comstock ore. This allowed little room for error. When all was said and done, the Washoe pan process simply left too much silver behind in the gangue.[21]

The company and Simeon Wenban spent the rest of 1864 in failed experiments aimed at improving the recovery rate at the Mill Canyon mill. A September 1864 article spoke of these difficulties, noting that "The ores, however, are generally sulphurets and consequently would require chlorinizing in order to properly work them. To do this, roasting furnaces must be built, and a material change made in the crushing and amalgamating machinery of the mill."[22]

Fortunately, "rebellious" ore was not unique to Cortez. The Austin mills had developed a process, naturally called the Reese River process, to deal with these same issues. The Cortez Company converted the mill at Mill Canyon to the Reese River process in 1865, with the addition of four reverbatory furnaces.[23] But the mill operators lacked experience with this method, and it was years before it operated efficiently with any consistency. One 1866 newspaper report attributed the district's problems to the fact that "the chief enterprise was in the hands of a San Francisco company, and has been so managed as not only to ruin nearly the whole company, but to depress every interest in the district."[24] In early 1867 the Tenabo Mill was described as "ill-arranged and poorly furnished."[25] Meanwhile, hauling ore to Austin by mule train for milling remained the only dependable, but costly, option for the Cortez mines.

Later in 1867, a newspaper wrote that the St. Louis mine had been "purchased by parties who are familiar with its character, and who are possessed of the capital, energy, and experience necessary to develop its treasure."[26] This may have referred to Wenban's purchase of controlling interest from George Hearst which, as we will see, was a crucial step in Wenban's takeover of the district. It was later reported that the Tenabo Mill was back, with repaired and improved machinery, "under the supervision of H. H. Day, formerly of the Savage Mine."[27] The Savage was one of the Comstock's best-known and most productive mines.

Many of the original Cortez investors and locators moved on to other opportunities during these first few years, when success looked anything but certain. By 1869, Andrew Veatch was working at the Erie Mine in California's Eureka District.[28] The original name for the Mill Canyon camp, Veatch's Camp, was dropped, and Veatch's Mountain became Mount Tenabo. Veatch would eventually be recognized as one of the foremost mining engineers in the region. John Cassell, whose 1873 obituary described him as "a well-known mining man,"[29] went to California to manage mining operations and practice law. Simeon Wenban not only stayed in the district, but as early as 1864 he had confidence enough in its future to at last send for his family to join him. His wife Caroline and their daughters Flora, age fourteen, and Eva, age twelve, journeyed from Ohio to Nevada by railroad, riverboat, and Overland Stage, riding the last leg of the trip from Grass Valley to Mill Canyon in an open wagon.[30]

The late 1860s and early 1870s saw two critical developments in the Cortez District, both tied closely to Simeon Wenban. First, he shifted mining operations from Mill Canyon to the Nevada Giant side of Mount Tenabo; and second, he consolidated all the major mines and claims under his own control. With the discovery and founding of the district during 1863 and 1864, the organizers and financial backers concentrated on developing Mill Canyon and putting the mill into operation. But the Mill Canyon mines quickly faltered, and the miners turned to the opposite side of the mountain.

In their first days on Mount Tenabo, the "expert of the company" (presumably Andrew Veatch) declared that silver would only be found in granite, not in limestone. He was only half right. They found silver in the granite in Mill Canyon, but he was wrong about the limestone. Wenban and others did explore the west side of the mountain and filed a number of "limestone" claims. They initially held off on developing them, but once they did it only took a few short years for the St. Louis, Arctic, Idaho, and Garrison mines to become the mainstays of the district. As the story goes, Wenban built a cabin for himself near the St. Louis mine, and dug out a quantity of rich ore and transported it to Austin by mule train. His profit after milling not only clothed and provisioned his family, but it confirmed his belief in the promise of Mount Tenabo's west side.[31]

Knowing the potential of the St. Louis, Wenban still had one obstacle to overcome. George Hearst had been among the early investors in the west side claims, and he still owned a controlling interest in the mine. Hearst's Comstock ventures had fallen on hard times, and he was either unable or unwilling to invest any more money in Cortez. Wenban traveled to San Francisco and with financial help from C. B. Land, his employer and associate from Virginia City, he bought out Hearst for $14,000.[32]

Wenban then went to work on the Garrison Mine, where he excavated enough ore to show it was even richer than the St. Louis. He bought up the last of the district's other claim holders and small investors, and from this point on he controlled all the significant mines in the district.

In 1869, Wenban bought the mill at Mill Canyon, which meant he could now process ore at his own facility. The exact details of the purchase are unknown, but Wenban acquired a mill that cost $100,000 to build for either $6,000 or $8,000 (sources differ). That may have been a fair price, considering the problems the mill was having and its remote location. On the other hand, Wenban apparently bought the mill from himself, since he was both buyer and an investor and official in the mill's owner—the Cortez Gold and Silver Mining Company and its successor, the Mount Tenabo Silver Mining Company, which operated the mill as late as 1868.[33]

In 1870, Arnold Hague, reporting on the Cortez District in the "Mining History" section of the *Geological Exploration of the Fortieth Parallel*,[34] wrote that in the past "conflicting interests, mismanagement in some cases, want of capital in others," along with high costs had "greatly retarded the development of this district." He said work had been suspended in the Mount Tenabo Company's Mill Canyon mine in 1868, and the district was "almost entirely deserted." By 1870, however, the issues had been resolved, and the district was "growing in importance."

Hague's report underscored the fact that by the late 1860s mining had shifted to the Nevada Giant side of Mount Tenabo. He described the St. Louis as the most prominent mine, with the Garrison showing considerable promise since it was opened in 1868. He also cited the Lander County Assessor's production figures for the last quarter of 1869, which listed four producing mines—the St. Louis Company, Mt. Tenabo (probably the Garrison), Berlin, and Arctic. All these were on the west side of Mount Tenabo. Ore previously sent to Austin for reduction was now being milled in Mill Canyon, although the mill's inconsistent performance meant the pack mule option was still utilized on occasion.

Wenban's consolidated ownership began fifteen years of steady growth for the Cortez District. Production figures showed increases from the early 1870s through 1885,[35] but there were also frustrations. He remodeled the Mill Canyon mill several times, with indifferent results, despite one newspaper article noting the success of the recently installed Bruckner furnace.[36] And there was the constant expense of either packing ore over or around Mount Tenabo to reach the Mill Canyon mill or hauling it all the way to Austin.

When the Nevada State Mineralogist summarized the situation in his 1873–1874 report, he said "success has crowned the mine owner after so many discouraging first attempts."[37] Production figures showed that between 1871 and 1878 the district produced between 340 tons of ore per year—gross yield $31,000 dollars—and 655 tons—gross yield $85,000. In 1879, production increased to more than 1,000 tons, valued at more than $87,000. Then tonnage remained near or above 1,000, reaching a high of 1,800 tons in 1884. Gross yield exceeded $100,000 in several years.[38] Newspapers, such as the *Sacramento Daily Record-Union* in 1875, reported Wenban's mines were turning a steady profit.[39]

Another government mining report in 1883 confirmed the Garrison as the Cortez District's chief producer. The Garrison tunnel penetrated 1,900 feet into the mountain, and additional drifts and cross-cuts aggregated to 9,000 feet (almost two miles) of underground workings. The report described the targeted ore body as "regular" (as opposed to irregular and difficult to follow through the parent rock) and increasing in size and grade. Ore was milled at Mill Canyon's 10 stamp mill, equipped with the Bruckner chloridizing furnace installed in 1876. It ran continuously and produced regular shipments of bullion. The report also mentioned two other lesser mines, the Silver Fleece and the Naiad Queen, neither of which Wenban owned. They included only a few hundred feet of underground workings, and the "rebellious" ore from the Silver Fleece had to be shipped to Salt Lake City or Eureka for smelting.[40]

Wenban's success in the 1870s and early 1880s laid the financial groundwork for the new Tenabo Mill and the expensive water supply system it would require. In 1886, with the rest of Nevada's mining industry sunk in *borrasca* (a Spanish mining term for a dearth of valuable minerals), Wenban and the Tenabo Mill and Mining Company built a new, state-of-the-art mill at Cortez. It was an ambitious undertaking and easily the most important investment to date in the district, if not the entire state at that time.

The Mill Canyon location made perfect sense as the site of the district's first mill. The mines were close by, and Mill Creek was one of the few substantial, reliable water sources in the area. But as mining shifted to the west side of the mountain, the location became more problematic. To get to Mill Canyon from the Nevada Giant side, ore had to be hauled four miles by mule train on trails over the flank of Mount Tenabo and down into Mill Canyon. Or it could be taken the eight miles by road out Cortez Canyon and around to Mill Canyon and the mill site.[41] Relocating the mill or building a new one closer to the mines was unworkable without an adequate water supply. For fifteen years, the alternatives were to get the ore to Mill Canyon one way

FIGURE 3.2. The Tenabo Mill, ca. 1888. The slope of Mount Tenabo is visible in the background. Courtesy of Northeastern Nevada Museum.

or another, take it to Austin, or in some cases take it to the railhead at Beowawe for shipping to more distant mills or smelters.

Wenban began planning for the new mill during the early 1880s. Work started on the water system as early as 1884, with construction of a pipeline from Wenban Spring to the future site of the Tenabo Mill.[42] The system included a siphon seven miles long that brought water from Wenban Spring on the west side of Grass Valley to the future Cortez townsite. Two wells on the valley floor, with steam powered pumps to push the water up the east slope of the valley, supplied the mill.[43]

Wenban built the Tenabo Mill close enough to the Garrison Mine that ore could be hauled by a tramway, or small railroad, directly from the mouth of the mine to the mill. The mill incorporated the latest technology and methods, including the new Russell lixiviation process for extracting silver from the ore. The Russell process was named for E. H. Russell, a Utah mine assayer. He developed his method in 1883–1884,[44] and the Bertrand Mill near Eureka was already using it successfully when Wenban built his mill. He was astute and knowledgeable enough to not only employ the new process but to eventually hire away the Bertrand Mill's manager.[45]

By eliminating extra transportation costs and employing the most efficient beneficiation methods, Wenban increased the profitability of his mines. This meant low-grade ore, which in the past had been discarded or left unexcavated, could now be mined and milled at a profit. The mines not only expanded into new areas of good ore, but they could now tap a range of other, previously unworked ore bodies.

The Tenabo Mill began operation in 1886, and total production in the Cortez District jumped to 5,813 tons from 1,652 tons of ore the year before.[46] The *Elko Free Press* reported that Wenban shipped 99 bars of bullion, valued at $121,000, during the two weeks between March 11 and April 7, 1886.[47] A newspaper article in January 1887 noted that Wenban "has been the sole owner of this valuable mining property for the past 30 years…and without outside aid has worked the mines from the grass roots, and is now turning out bullion at the rate of $75,000 per month."[48] Production continued at this level, with the exception of one downturn in 1892–1893, until 1904. The best years were 1887 to 1891, when the district averaged more than a third of a million dollars a year.[49]

The water supply system that served the Tenabo Mill also provided water for a small nearby community. The new town of Cortez, with its proximity to the mines and mill, quickly drew residents away from Mill Canyon and Shoshone Wells. Cortez eventually included a hotel, boardinghouse, and company store, and had stage lines linking it to Austin and Beowawe. While no more than a few hundred people ever made it their home, it was the district's population center from the mid-1880s to the early 1940s.

The Cortez District eventually attracted the attention of British investors. The British were significant players in the North American mining industry during the last quarter of the nineteenth century. Wenban's English background also probably appealed to these potential associates, and he received an offer, bonded for $1,300,000, for his Cortez properties. He reacted coolly, at least in public. A newspaper article described him as just as happy to keep his mines, since "there is more ore in sight in them today than ever."[50] Nevertheless, in 1887 and 1888 Wenban visited England and negotiated the formation of the Tenabo Mill and Mining Company, to be operated by the Bewick-Moering Syndicate under the name Cortez Mines, Ltd. Bewick-Moering was one of the largest and most prestigious mining investment firms in the world. Wenban retained the majority of shares in the new company and served as its first managing director, at a salary of £2,400 British pounds per year (about $4,000 US dollars at the time).[51] In 1890, the Cortez property was described as the most valuable in the state. Assets included a mill, assay office, blacksmith shop, three large stables, store, and hotel. It employed more than 100 men and returned a profit of $30,000 per month.[52]

Wenban continued improving and expanding the mill, and he opened two new mine tunnels to reach ore bodies deep within Mount Tenabo. The Arctic tunnel began some 300 feet vertically below the mouth of the Garrison, aimed at intercepting ore exposed in the Garrison's lowest levels. A second tunnel extended over 4,000 feet to the St. Louis mine and tapped rich, deep ore Wenban had first mined in 1867 and 1868.[53]

Government monetary policy caused a collapse in silver prices in 1892, and the subsequent Panic of 1893 brought on the worst economic depression to date in the country's history. Concern grew in Cortez that low silver prices would cause Wenban to close the mines and mill and, as feared, by the end of 1892 Cortez joined other districts, including Candelaria, Tuscarora, Galena, and Grantsville, and ceased operations. Production in the entire Cortez District fell to a mere 36 tons of ore in 1893.[54]

Bewick-Moering turned the Cortez properties back to Wenban, and operations stayed at minimal levels for the next two years. But the district recovered, and the company was able to refurbish the mill in the late 1890s. The Cortez District recorded nearly one million dollars in total production between 1897 and 1902, with production returning to a pre-panic level of 5,900 tons by 1897.[55] Wenban, as he had for almost three decades, remained the single most important individual in the district, and he was rewarded accordingly. He earned a salary of $1,000 per month as president of the Tenabo Mill and Mining Company, along with whatever profits his stock ownership netted him.[56] In March, 1901, Wenban suffered an asthma-induced heart attack while overseeing work on his Cortez properties. He was taken immediately to San Francisco but died at his home a few days later. He was seventy-seven years old and left behind his widow Caroline, his two daughters, and at least four grandchildren.[57]

The Cortez District had been good to Wenban and his family. He left an estate valued at $567,000, including 252,470 shares of stock in the Tenabo Mill and Mining Company, valued at $260,000. His San Francisco properties comprised the Wenban Hotel, the family home on Van Ness Avenue, and other real estate.[58] The Wenban estate continued the mining operations as the Cortez Mill and Mining Company, but silver prices again fell in the years after Wenban's death. In mid-1904, work at the mines and the Tenabo Mill essentially came to a stop. Early that year, the company payroll listed thirty or forty mine workers and approximately the same number in the mill. By July, only a handful of employees were left.[59] When mining did resume in the district it was under leasing contracts between the Wenban estate and a long list of small operators, such as Cortez Leasing Company, Cortez Metals Recovery, or Cortez Mining and Reduction.

The first decade of the twentieth century saw what historians describe as Nevada's second mining boom. This time the southern part of the state took the lead, with discoveries in the Tonopah, Goldfield, and Bullfrog Districts. Occasionally, the rush to these districts left places like Cortez nearly abandoned. The cyanide leaching process and other advances in beneficiation motivated much of the boom. This process used a cyanide solution to leach gold and silver from the pulverized ore, after which the solution was filtered and processed to extract the precious metals. The first cyanide mills were built in the early 1890s, and within a few years the method dominated milling in Nevada. It was so efficient at recovering gold and silver that mills began processing their old tailings as well as low-grade ore that previously

FIGURE 3.3. The Wenban home on Van Ness Avenue in San Francisco.
Courtesy of San Francisco History Center, San Francisco Public Library.

went untouched. In some of Nevada's dormant districts, companies set up small cyanide mills and "mined" the old tailings, without ever going into a shaft or tunnel. At Cortez, the Cortez Metals Recovery Company converted the Tenabo Mill to cyanide leaching in 1908. The company engineer, A. W. Geiger, reported in an article on his work that "Owing to faulty [previous] lixiviation washing methods, about half the metal content of the tailing was readily soluble in weak cyanide solution, enough more to make up an 80 percent recovery."[60] Cyanide reprocessing continued at Cortez until the Tenabo Mill burned in 1915.

During the 1910s, the Cortez Mill and Mining Company continued giving way to smaller independent operators who leased various claims from the company. Their work was sporadic and disorganized compared with the decades of operation under Wenban's supervision. Lessees only went after what they thought would be the most profitable claims, or the most profitable areas within existing mines. By 1920, the population of Cortez had dropped to thirty-five people, with only two families making a living from mining. The once thriving mining camp was reduced to a handful of households clustered around the company store and boardinghouse, post office, and school.

Some work also shifted during this time to Mill Canyon and the surrounding mountains. The Buckhorn District, about seven miles from Cortez in the southern foothills of the Cortez Mountains, flourished briefly around 1910. In Mill Canyon proper, the Cortez Mining and Reduction Company mill, headed by John Blair Menardi of Goldfield fame, operated a mill between 1913 and 1915 just downstream

FIGURE 3.4. Rob McQueen recording the Menardi Mill, 2013. Courtesy of Summit Envirosolutions, Inc.

from the original Mill Canyon mill. The Menardi Mill was a 50-ton cyanide plant that produced concentrates that were trucked to Beowawe for shipment to smelters in Utah.[61] It burned in 1915 and was not rebuilt.

But change was on its way to the west side of Mount Tenabo. A group of investors had been scouting the Nevada Giant mines, and in 1919 they incorporated the Consolidated Cortez Company.[62] They bought the Cortez Mill and Mining Company, and all its mining property, from the Wenban estate. The new company's main office was in Reno, but it had national backing. Other corporate addresses were on Wall Street and in the prestigious Call Building in San Francisco. Its directors resided in New York, San Francisco, Reno, and Ashville, North Carolina.[63] Consolidated Cortez took an informed, calculated risk, similar to the original Veatch party who, it could be said, made their own luck. Consolidated Cortez's financial experts, geologists, and mining engineers were what we would recognize today as a corporate team, not Wenban's one-man rule or the lone prospectors and lessees who worked the district after his death.

The *Mines Handbook* of 1922 described the Consolidated Cortez management as unusually competent.[64] The company president, Frank Manson, was a seasoned mining man, with experience as head of the Western Ore Purchasing Co. and chairman of the Rochester Nevada Silver Mines Co. But even the best planned mining venture was a gamble. One financial guide, the *United States Investor* from March 1920, cautioned a reader who wrote in asking whether "these 'Cortez' mines are nearly exhausted and that it is only the high price of silver which makes them worthwhile?"

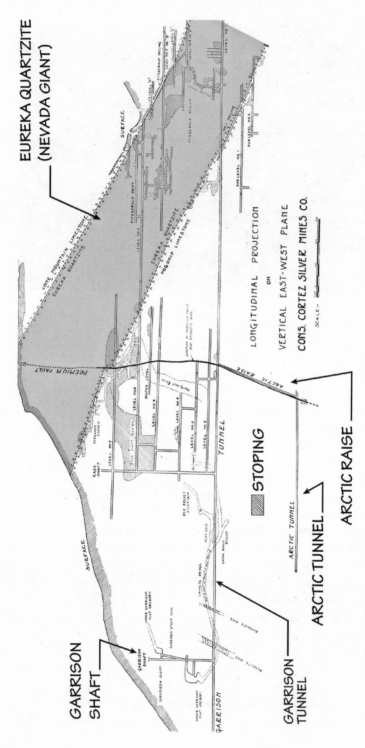

FIGURE 3.5. Cross-section of Mount Tenabo showing the extensive underground workings dating from the Wenban era to the early 1920s. Courtesy of Special Collections, University of Nevada, Reno Libraries. Tangerine Design & Web.

The response concluded that mining stocks were "not very far removed in specula-tive risk from the offerings of oil promotion" and said "We are not impressed with the likelihood of this venture working out to the expectations of the promoters. It seems that if the ore value of the properties controlled by the Cortez Silver Mines were as valuable as claimed…the mines should not have been abandoned."[65]

Consolidated Cortez had a different opinion. The company hired a number of prominent mining geologists to inspect the mines, in particular the Garrison and the Arctic. They reported back most favorably, and their endorsements were assem-bled in a prospectus published by Consolidated Cortez. It aimed, of course, to per-suade potential investors to buy stock in the company, but the experts shared the opinion that Mount Tenabo still held significant, unexcavated ore bodies accessible from the Garrison and the Arctic tunnels. One consultant, basing his calculations on underground ore bodies either in sight or well exposed, estimated potential profits at $778,434.[66] Net value of production from 1919 to 1929 turned out to be $2,036,892.[67] In this case, reality surpassed even the most optimistic projection.

Consultants C. K. Clark and J. H. Kramer reported, "We know of no property for sale in the State of Nevada that offers the opportunity for profitable investment as do the mines of Tenabo Mill and Mining Company. With the immense bodies of low and medium-grade ore already in sight and partially or wholly developed, the element of chance would seem to be entirely eliminated."[68]

In the words of the prospectus, the company possessed "a huge reserve of good milling-grade ore." They re-equipped the mines and began excavating. By 1923, after stockpiling enough ore to last several years, they built a state of the art concentration-cyanidation mill at the mouth of Arctic Canyon. The Consolidated Cortez Mill was described as "the most modern ore reduction plant in the West," and an "ingenious" combination of modern concentrating and cyaniding methods. Its recovery rate for silver, and also gold and lead, exceeded 90 percent.[69]

Consolidated Cortez began their mining deep within the mountain by expand-ing the Garrison's main tunnel, which had been the district's top producer going back to 1880 and before. The work included networks of drifts, cross-cuts, winzes, and raises. The Arctic tunnel was driven over 2,500 feet into the interior of the mountain to open access to the ore bodies between the two mines.[70]

The Consolidated Cortez Mill represented another step forward in milling tech-nology and significantly improved the recovery rate for silver from the Cortez Dis-trict. Like the Tenabo Mill before it, this opened up new areas for mining and made previously excavated low-grade ore profitable. Much of the projected value was in dumps from earlier mines and from stope-filling. Miners from the previous cen-tury had dug out this low-grade ore but left it underground in empty stopes and tunnels rather than expending time and money transporting it to the surface. Now that it could be profitably milled, it was loaded into ore cars and sent on its way to the mill.[71]

New ore came from the lower levels between the Arctic and the Garrison. Employing a technique called "shrinkage stoping," miners worked their way upward into the ore bodies from the Arctic tunnel. They began by excavating a drift below the targeted area, then dug a series of short shafts, or raises, vertically into the bottom of the ore body. These raises became chutes through which broken rock fell into ore cars waiting below on rails installed in the drift. As the miners expanded their excavation upward and outward from the raises—forming a stope—the chutes could be closed off to allow broken rock to accumulate and form a platform under the feet of the miners as they drilled and blasted into the roof. Ore was then released through the chutes at the same rate newly mined rock accumulated, allowing the floor of the work area to keep pace as the stope expanded upward. Once the miners reached the top of the ore body, they continued emptying the stope until all the ore was gone.

The Consolidated Cortez Mill operated as a cyanide reduction mill for four years. In 1927, the company converted it into a 150-ton flotation mill.[72] Flotation was another new beneficiation process, in which a slurry of crushed ore and selected oils and other liquids was injected with compressed air. Particles of precious metal adhered to the air bubbles as they rose, forming a mineral-rich froth, which was skimmed off and sent to the refinery.[73]

The 1920s saw production in the Cortez District equaling its best years from the previous century. A 1930 economic report on the Consolidated Cortez Company stated that from the purchase of Tenabo Mill and Mines Company in 1919 through 1929 the District produced 267,249 tons of ore with a net (after expenses) value of $2,036,892.[74] In 1928, the Cortez Mining District was Nevada's leading silver producer. In 1929, the Consolidated Cortez Mill employed sixty workers and processed 125 tons of ore per day. The ore was reduced to concentrate and trucked north through Crescent Valley to the railhead at Beowawe, then shipped by rail to a smelter in Salt Lake City for final refining.[75]

The price of silver began falling precipitously in the late 1920s, first in response to excess supply and declining demand on international silver markets and then to the stock market crash and the onset of the Great Depression. The price of silver had fluctuated between 53 and 60 cents per ounce from 1926 to 1929, but it declined steadily through 1929 and 1930. It would reach an all-time low of 25¾ cents per ounce in February 1931.[76] The Consolidated Cortez Mill was only breaking even, and after a number of failed attempts at staying in production the company closed the mill in 1930.[77]

With work stopped on their properties, Consolidated Cortez eventually went into receivership in 1937. A New York firm acquired the assets, and it re-emerged as the Cortez Metals Company. However, lawsuits tied up the company's properties for the remainder of the decade. The Consolidated Cortez Mill was reconditioned in 1939 with new flotation equipment, but this revival was short-lived.[78]

FIGURE 3.6. The Roberts Mill ruins. Courtesy of Summit Envirosolutions, Inc.

The 1930s did see a shift in mining and milling back to the Mill Canyon side of Mount Tenabo. In 1937, the Roberts Mining and Milling Company completed construction of a 150-ton capacity mercuric cyanide mill in Mill Canyon. They started the mill in 1930, but litigation, financial problems, and open conflict among the principals, including a dynamited ore bin, kept it idle for many years.

In September 1937 the Roberts Mill began working ore from the Emma E. Mine in Mill Canyon. An aerial tramway conveyed the ore along a ridge from the mine to a road, where it was trucked 1.3 miles down the canyon to the mill. New lawsuits again suspended work until the summer of 1938, when the former Consolidated Cortez manager leased the Roberts property and resumed mining. The mill closed once and for all in April 1940, as the Emma E. exhausted its ore.[79]

The US government ended precious metal mining during World War II by closing mines not engaged in producing strategic metals and minerals. The Roberts Mining and Milling Company was bankrupt by 1945.[80] A few more failed attempts followed after the war, but mining and milling in the Cortez District was essentially over. In 1954, the Consolidated Cortez mill was dismantled and scavenged for salvageable materials.

Nevada historical writer Nell Murbarger chronicled the last few years at Cortez, summarizing her visits in a 1963 article for *True West*. She wrote that during the 1950s

FIGURE 3.7. Lloyd High, last resident of Cortez. Courtesy
of True West Magazine.

a dozen or so people still made their homes in Cortez.[81] The towering brick chimney at the old Tenabo Mill still stood, along with the big company office and boardinghouse. But when she visited in 1963, the only person there was Lloyd High, the
self-appointed caretaker of both the town's abandoned structures and its word-of-
mouth history.

Lloyd was a colorful figure who enlivened his solitude with poetry, songs, and
a guitar. He grew up in Montana, served in WWI, had a ranch for a time in Grass
Valley, and then took up mining at Cortez in 1919.[82] He was in awe of the history
around him and wished he could have lived at Cortez in Wenban's day when, according to the old timers, it was a very busy place. He told Nell Murbarger, "It's almost
impossible to imagine the amount of work that has been done here."[83] He described
how, along with miles of tunnels, shafts, and stopes hidden within Mount Tenabo,
people made their own bricks for building the mill, manufactured their own lime,
gathered salt for processing ore, and toiled day after day in the mills. As for modern
Cortez, he wrote himself that "Cortez lies sleeping on the breast of Mt. Tenabo,
among the pinon pines, junipers and sage brush where hundreds of charcoal burners,
miners, muckers, mill men, woodchoppers and longline skinners once made a living
and raised families."[84]

NOTES

1. Estimates vary for total historic production (1863 to ca. 1950) for the Cortez District. A compilation from Vandenburg (1938, 23) indicates production between 1863 and 1903 totaled about $10,000,000, with an additional $2,700,000 worth of silver and gold between 1903 to 1936.

2. Ligenfelter (1974, 107–127); Raymond (1869, 3–6); Wyman (1979, 37–41). More specifically, Wenban *successfully* employed Chinese hardrock (underground) miners and millers. Other mining camps in the West (California and Nevada especially, but also Colorado and Idaho), attempted to hire the Chinese, but powerful miners' unions, threatened by the lower-wage workers, were effective in halting the practice. Wenban's use of Chinese miners lasted approximately twenty to twenty-five years. In comparison, Chinese miners at Silver Peak, Nevada, were fired after only one year. The lengthy employment of Chinese miners at Cortez might also hint at the lack of organized union activity in the district, at least under Wenban's watch.

3. Hess et al. (1987, 9); Procter (1998, 6); DeQuille (1877, 60).

4. *Reese River Reveille*, October 28, 1863, 1.

5. *Reese River Reveille*, October 28, 1863, 1; *Reese River Reveille*, July 25, 1863.

6. Simpson (1876).

7. Beckwith (1855).

8. Bancroft (1889, 248): "But the expert of the company had an opinion, to which he adhered, that silver was only found in granite." The "expert" was presumably Veatch. Wenban held his peace for the time being, but he eventually joined in making important discoveries in the limestone on Mount Tenabo's west slope.

9. *Reese River Reveille*, June 27, 1863.

10. Bancroft (1889, 248); Cortez Mining District (n.d.).

11. *Reese River Reveille*, May 5, 1864, 1: "Mssrs. Cassell, Wenban, and their associates, after locating a large number of ledges in bullion and Little Bullion Hills, completely covering both, formed, with a number of San Francisco capitalists, the Cortez G. and S. M. Company."

12. Knight (1864). Knight edited *Bancroft's Hand-book Almanac for the Pacific States*, which described these notable figures.

13. We know few details of Hearst's actual participation and investment in the Cortez District. According to Bancroft (1889, 250), Wenban bought out Hearst for $14,000, which he obtained from one of the Land brothers. Hearst's initial investment is unknown, but he was listed about this time as a delinquent shareholder in the St. Louis Mine. He was a well-known mining figure and a major stockholder in at least one of the Cortez properties. Buying him out was critical to Wenban's ultimate control of the District.

14. *Reese River Reveille*, August 8, 1863.

15. See Introduction, note 4.

16. *Reese River Reveille*, September 5, 1863, 3. The notice is dated August 24, 1863, also the date of the meeting and assessment.

17. *Reese River Reveille*, December 15, 1863, 2.

18. *Reese River Reveille*, May 5, 1864, 1.

19. Bancroft (1889); *Reese River Reveille*, May 7, 1864, 1.

20. *Reese River Reveille* May 7, 1864, 1: "The driving shaft is solidly placed in frame by itself, connected with rubber belting to the battery and pan shafts, thus relieving it from the jar and irregular strains a more direct connection would have caused."

21. Bancroft (1889, 248–249); *Reese River Reveille*, September 28, 1864.

22. *Daily Reese River Reveille*, September 28, 1864.

23. Bancroft (1889, 249).

24. *Daily Reese River Reveille*, February 20, 1866.

25. *Daily Reese River Reveille*, January 3, 1867.

26. *Daily Reese River Reveille*, June 7, 1867.

27. *The Anglo American Times*, January 25, 1868, 8.

28. *The Sacramento Daily Union*, September 17, 1869.

29. *San Francisco Call*, July 3, 1893.

30. Magee (2010, 23–43).

31. Bancroft (1889, 250).

32. Bancroft (1889, 250).

33. There are no original records of the sale, however secondary sources (Angel 1881, 429; Hezzelwood 1930, 2; Hardesty 1988, 83) all place it in 1869. Wenban bought the mill for much less than its $100,000 construction cost, perhaps for as little as $6,000 (Bancroft 1889, 252; Angel 1881, 429).

34. Hague (1870, 405–407).

35. Couch and Carpenter (1943, 58–59).

36. *Sacramento Daily Record*, August 4, 1876.

37. *Biennial Report of the State Mineralogist of the State of Nevada: For the Years 1873 and 1874.*

38. Couch and Carpenter (1943, 58–59).

39. *Sacramento Daily Record-Union*, March 13, 1875, described the Garrison as a bonanza; *Sacramento Daily Record-Union*, April 12, 1875: "Wenban Mine at Cortez is turning out twenty-five tons of ore per day @ $150 per ton."

40. United States Bureau of the Mint (1883, 151).

41. Angel (1881, 429) provides mileages and notes that the ore was carried by pack mules. Modern maps show the trip out Cortez Canyon and up Mill Canyon was closer to ten miles. The more direct trail over Mount Tenabo was about four miles, but meant climbing 2,000 feet up the mountain and then dropping down 2,000 feet into the canyon.

42. *Daily Nevada State Journal*, June 11, 1884: "At Cortez Mr. Wenban is building a new mill and laying a pipe to bring water in from the mountains."

43. Hezzelwood (1930, 12). He observed in 1930 that the Cortez camp was supplied by gravity flow from a spring seven miles south, while "Water for the mine and mill is pumped against a 500-foot head from wells in the valley." We were unable to confirm this separation during our investigation.

44. Stetefeldt (1888, 5, 7): "The Russell process was first introduced in September, 1884, at Silver Reef, Utah… At the Ontario Mill, Park City, Utah, Mr. Russell experimented on a large scale in 1883–4." Referring to the slightly earlier history of lixiviation in Nevada, Stetefeldt (2) says: "The use of the process was revived by the construction of the Bertrand Mill, in 1882, the Mount Cory Mill, 1883, and the Cortez Mill, in 1885, all in Nevada."

45. Johnson and McQueen (2016, chapter 95: Mining and Milling Technology).

46. Couch and Carpenter (1943, 58–59).

47. *The Free Press*, May 14, 1887.

48. *White Pine News*, January 8, 1887, 3.

49. Couch and Carpenter (1943, 58–59).

50. *Reese River Reveille*, February 6, 1887.

51. Spence (2000,118). The Consolidated Cortez Silver Mines Company prospectus (page 4), noted Cortez Mines, Ltd. was capitalized for 300,000 British pounds, par value 1 British pound per share, with 290,000 shares issued to the Wenbans.

52. *White Pine News*, December 27, 1890.

53. Consolidated Cortez Silver Mines Company (n.d., 4).

54. *Nevada State Journal*, April 10, 1892; May 7, 1892. Newspapers begin mentioning problems with the Cortez mines in 1891. *The White Pine News*, May 4, 1891, noted that "the Cortez mining camps is no place for miners to seek work." The same newspaper wrote on April 23, 1892, that the Cortez mill was closed owing to the low price of silver and that people were thoroughly discouraged. According to Consolidated Cortez Silver Mines Company (n.d., 4): "Disputes over the finances of the corporation in the early part of 1892 resulted in the suspension of operations and the Wenbans and the Bewick-Moering Syndicate members spent a great part of their time in court."

55. Couch and Carpenter (1943, 58–59).

56. Tenabo Mill and Mining Company Payroll Ledgers, February, March, May–December, 1897.

57. *Reno Evening Gazette*, March 5, 1901, 1: "Simeon Wenban Dies in San Francisco"; *San Francisco Call*, March 5, 1901: "Simeon Wenban, Pioneer Mining Man, Passes Away."

58. *San Francisco Call*, March 13, 1901, 7: "Simeon Wenban's Will."

59. Tenabo Mill and Mining Company Payroll Ledgers, June 1896–March 1908.

60. Emmons (1910, 101): "In the summer of 1908 cyanide tanks were built to treat the tailings from the Cortez mill, which are estimated to amount to about 120,000 tons. At this time the mines were not producing." A. W. Geiger's comments were in a letter published in *Mining and Scientific Press*, December 14, 1912, 767.

61. *Nevada State Journal*, October 17, 1913, 8.

62. Consolidated Cortez Silver Mines Company (n.d., 1).

63. Consolidated Cortez Silver Mines Company (n.d., 1).

64. Weed (1922, 1173–1174).

65. Bennet (1920, 24g–24h).

66. Consolidated Cortez Silver Mines Company (n.d., 11).

67. Hezzelwood (1930, 2).

68. Consolidated Cortez Silver Mines Company (n.d., 11).

69. Consolidated Cortez Silver Mines Company (n.d., 5).

70. Consolidated Cortez Silver Mines Company (n.d., 5).

71. Consolidated Cortez Silver Mines Company (n.d., 10); *Mining and Scientific Press*, May 29, 1920, 804.

72. Vanderburg (1938).

73. Bunyak (1998).

74. Hezzelwood (1930, 2).

75. Johnson and McQueen (2016, chapter 95: Mining and Milling Technology).

76. Dickson (1939).

77. *Reno Evening Gazette*, October 7, 1933, 7; Hardesty 2010.

78. Johnson and McQueen (2016, chapter 95: Mining and Milling Technology, citing *Engineering and Mining Journal* 1939, 68); *Nevada State Journal*, August 8, 1938, 5; Vanderburg (1938, 22).

79. Johnson and McQueen (2016, chapter 95: Mining and Milling Technology, 15, citing Bertrand 1993, 86); *Reno Evening Gazette*, September 1940, 10; *Reno Evening Gazette*, October 7, 1944, 8; McQueen et al. (2015, 10–13).

80. *Reno Evening Gazette*, June 5, 1945.

81. Murbarger (1963).

82. *Nevada State Journal*, June 24, 1956.

83. Murbarger (1959, 13).

84. High (n.d., 9).

Chapter 4

Unlocking the Silver

During the Cortez District's long lifetime, the mining world saw one advance after another in beneficiation—the art and science of extracting maximum value from silver ore. Most all of these advances found their way to Cortez, sometimes as near-desperate measures to keep the district afloat and sometimes as well-planned, well-funded implementation of the best, most up-to-date methods available.[1]

THE WASHOE PAN PROCESS

When the Mill Canyon mill went into operation in 1864, it extracted silver from the Cortez ore using the Washoe pan process, named for the Washoe District where it was widely used. It was an elaboration upon the simpler *patio process*, in which finely pulverized rock was mixed with water, salt, mercury, and copper sulfate and then spread over a rock pavement, or patio. The sun's heat and constant mixing caused the silver compounds and mercury to amalgamate, or bond together. The amalgam was then separated from the mix and further processed to remove the gold and silver. In the Washoe pan process, heavy metal tubs or pans, such as the Knox amalgamating pans found at Cortez, replaced the open air patio.

At Mill Canyon, the ore was first fed into a coarse crusher and went from there to the stamp battery. The battery was comprised of eight 700-pound stamps—large, crushing hammers that repeatedly pounded the ore until it was reduced to powder. The crushed ore was then mixed with water and other chemicals, mercury in particular, to form a *pulp*, which flowed to settling tanks. From there, the settled pulp was loaded into cars that ran on a track between the rows of amalgamating pans. There were twelve Knox amalgamating pans, about five feet wide and one foot deep. Each one held 1,200 to 1,800 pounds of pulp, heated by steam introduced into chambers beneath the pans. When the amalgamation process was complete, the pulp was transferred to large wooden tubs, or settlers. These were about twelve feet in diameter, and stirred by rotating arms. The heavy mercury-silver amalgam settled to the bottom and was removed to the combination retort and melting furnace. Here the mercury vaporized and was captured in a condenser for reuse. The remaining amalgam was melted and poured into bars of silver bullion.[2]

THE REESE RIVER PROCESS

The Washoe pan process proved a failure with the "rebellious" Cortez ore, and the Cortez Company converted the Mill Canyon mill to the Reese River process in 1865, with the installation of four reverbatory furnaces. In the Reese River process "the ore is dried before stamping, stamped dry, and roasted with salt before being treated in the pans."[3] Salt was added to the ore as it underwent final crushing in the stamp battery, and then it went to the furnaces for up to seven hours of roasting. "The silver minerals [reacting with the chlorine in the salt] are converted to chlorides, which amalgamate easily."[4] Antimony and arsenic, and the troublesome base metals—lead and arsenic—also form chlorides and at least partially volatize and escape up the chimney.[5]

Manuel Eissler explained the process in *The Metallurgy of Silver*, a book on milling methods of the day:

> [Rebellious] ores as cannot be treated by the Washoe process on account of the quantity of base metals they contain have to be roasted before being subjected to amalgamation. The object of the roasting is not only to drive away the sulphur, antimony, arsenic, and other volatile products, and set the silver free from these combinations, but by the addition of salt to convert the base metals as well as the silver into chlorides, which, in the subsequent amalgamation, are reduced and combine with the quicksilver to form an amalgam.[6]

The 1870 report on the Cortez District in the *Geological Exploration of the Fortieth Parallel* by Arnold Hague gave the full lineup of machinery in place in the Mill Canyon mill. He wrote that the mill "is provided with 12 stamps, 4 roasting furnaces, 4 Varney Pans, and 2 settlers. The reduction process is the same as that employed at Austin."[7]

The Wheeler and Varney amalgamating pans were an additional improvement. These were similar pieces of machinery, both designed to enhance the amalgamation process by grinding the liquefied pulp and thoroughly mixing it with mercury. An ordinary Wheeler pan was about four feet in diameter and two feet deep. The muller, mounted on a rotating shaft, ground the pulp against the flat-bottomed pan and also kept the mix in continuous circulation to maximize exposure of the ore particles to the mercury. Steam chambers at the bottom of the pan provided heat for the process.[8] The pulp went next to the settlers, where—just as before—the heavy amalgam settled out and was taken to the retort furnace.

THE RUSSELL PROCESS

Simeon Wenban installed the Russell process at his new Tenabo Mill in 1886. This was state-of-the-art beneficiation, with improved roasting and chloridizing and new chemistry to replace mercury amalgamation. The Russell process was a "lixiviation" or leaching process in which crushed ore was saturated with a chemical solution that bonded with the precious metal as it percolated through. The solution was then recaptured and further treated to precipitate the silver.[9] H. H. Bancroft described the

FIGURE 4.1. Bruckner's roasting cylinders at the Tenabo Mill. Summit Envirosolutions, Inc. Photo by Robert McQueen.

workings of the Russell process in detail in his 1889 biography of Simeon Wenban: "The ores being chiefly rebellious, were put through a variety of processes, going from a drying furnace hot to a rock-breaker, thence to crushing rolls and revolving screens, and all automatically."[10]

The Tenabo Mill utilized Krom's rollers instead of a stamp battery.[11] Krom's rollers crushed ore between two heavy, rotating cylinders, rather than pounding it with steel hammers. They were a major improvement over stamp batteries and were a further indication of Wenban's commitment to the latest technologies. Krom rollers were cheaper to buy and maintain than stamps. The steel "tires" that were the actual crushing surface lasted longer than the steel hammer and anvil. At the Bertrand Mill (near Eureka), they anticipated crushing 20,000 tons of ore in their Krom rollers before they would need replacing. The rollers crushed ore faster than stamps, because they subjected the material to continuous crushing action rather than a series of separate blows. Unlike stamp mills, which produced so much noise and dust they often had to be isolated in their own buildings, Krom rollers could be housed with other machinery in the main mill.

A book on the subject quoted Simeon Wenban's own endorsement of Krom rollers. He said, "I have been an advocate of stamps, but after seeing the quantity of ore your rollers have crushed at the Bertrand Mill (30,000 tons), I am convinced of their superiority over stamps, and have decided to use *them* in my new mill."[12] Continuing his description, Bancroft wrote: "Then followed the chloridizing furnace, where the heat drove off the bases in fumes, and a chloride was formed with common salt." The chloridizing furnace was comprised of two rotating Bruckner cylinders. These were built in Cincinnati by Lane and Bodely Company and shipped to Cortez.[13]

Bancroft concluded: "from the chloridizing furnaces to the leaching vats, where it is first treated to a water leach from two to three hours; afterwards the leaching process is continued with a solution of hyposulphite and bicarbonate of soda, which dissolves the chloride of silver, and passes from the leaching vats into precipitating tanks, into which is introduced a quantity of lime and sulphur solution; this precipitates the silver which immediately settles to the bottom of the tanks. Soon as the silver (now in the condition of a sulphide) is sufficiently settled, the hyposulphite and bicarbonate solution is drawn off and pumped back to the storage tanks to be used over again—the silver sulphide is taken from the tanks, forced through a filter press, from whence it comes in cakes; these are passed to a drying-room, from there to a furnace which frees it from the sulphur, and is then ready to be melted into bars for shipment."

CYANIDE LEACHING

Cyanide leaching was a lixiviation method in which ore was saturated with a weak cyanide solution to dissolve the gold and silver, which was then recovered from the solution once it drained away. It represented a major improvement in recovery rates, to the point where it was commonly used to reprocess mill tailings and salvage gold and silver the older processes left behind. Cyanide leaching was introduced at Cortez in 1908, and it was used on tailings from the Tenabo Mill. After recrushing, the tailings were mixed with cyanide solution in large wooden tanks. The resulting slurry was filtered and the "pregnant" solution—still containing dissolved metals—drained off. Zinc shavings were then added to precipitate the gold and silver.

Cyanide leaching was an important part of the operation implemented by Consolidated Cortez in 1923. They described their mill as "ingeniously combining standard concentrating and cyaniding methods into one efficient process."[14] The ore was fed into a crusher, size sorted on screens, and conveyed to a ball mill and tube mill where it was ground to powder. It was mixed with cyanide solution in thickening and agitation tanks. The precious metals were precipitated and then extracted in two Oliver filters and then sent to the furnace to be melted and poured into bullion.[15]

FLOTATION

The Consolidated Cortez Mill was remodeled for flotation in 1927.[16] Flotation was a successful but temperamental process that seemed just as much art as science. It exploited differences in specific gravity to separate precious metals from the gangue. To begin, the pulverized ore was mixed with various oils and chemical reagents.

A stream of compressed air was injected into the slurry and the mineral particles adhered to the rising bubbles while the gangue slid off and sank. The froth was then skimmed and sent on for further processing. Flotation required experimentation and constant adjustment, as chemistry varied even within a single ore body, and the relationship between the specific gravity of the mineral and the surface tension of the bubbles had to be perfect. Pine oil was the preferred medium at Cortez, but the list of oils employed in flotation schemes included coal tar, creosote, kerosene, and petroleum. Millers also experimented with fish, whale, corn, cotton seed, and rape-seed oils, not to mention lard, turpentine, and even sagebrush oil.[17]

NOTES

1. Eissler (1891).
2. *Reese River Reveille*, May 7, 1864.
3. Emmons and Calkins (1913, 195).
4. Emmons and Calkins (1913, 195).
5. Eissler (1891, 147).
6. Eissler (1891, 145).
7. Hague (1870, 406).
8. Eissler (1891, 62–66).
9. Eissler (1891, chapter 13).
10. Bancroft (1889, 256).
11. Hardesty (1988, 83, citing Bancroft 1889 and also *Mining and Scientific Press*, May 29, 1920, 803–804); Eissler (1891).
12. Krom (1885, 18); *Engineering and Mining Journal*, July 11, 1885, 27.
13. Johnson and McQueen (2016, chapter 95: Mining and Milling Technology); *Engineering and Mining Journal* (1885:76); Eissler (1891, chapter 9).
14. Consolidated Cortez Silver Mines Company (n.d., 5).
15. Hardesty (1988, 58–59).
16. Vanderburg (1938).
17. Megraw (1918, 56–68).

Chapter 5

A Ton of Work for an Ounce of Silver

From Mount Tenabo, in the late 1880s, you would find the slopes and foothills below you alive with activity. Miners would work invisibly deep beneath your feet, although you might sense a slight tremor as they blasted loose another ton of rock. The rumble from the crushers at the Tenabo Mill would be no more than a distant murmur. The smoke pouring from the towering brick stack would be black against the sky. On the mountain, woodcutters would chop at the pinyon and juniper. Charcoal makers would load and stack logs while wisps of smoke rose from their mounded, earth-covered charcoal ovens. Somewhere else, lime makers would tend their fiery, cylindrical kilns. Farther away on the valley floor brickmakers would be mixing clay, filling molds, and firing grand stacks of brick. And beyond that, on the shimmering alkali playa, the tiny, dark dots would be people scraping up salt.

The Cortez District was a major industrial operation, created from whole cloth in the middle of nowhere. It could only succeed if each ounce of silver was sold for more than it cost to produce. That goal came closer every time the people of the district found, made, or created something instead of buying it and paying to have it brought to them. Still, they depended on the outside world for the machinery they could not dream of making themselves, the food they ate, and their letters from home. In return, they sent silver to that same world to be coined, spent, or hoarded in places as familiar as Carson City or San Francisco or as foreign as London or China.

An archaeologist sees this as one system, its parts all interrelated and all working toward a common objective. At the core are mining and milling, producing silver bullion at a profit. But where did the energy to do this come from? What about the chemical ingredients vital to beneficiation? Building materials? Water, not only for the mill, but to sustain life itself? And finally, how did it all get to the right place in the system at the right time? How did the district obtain what it needed from the outside world, and how did it move its product to that same world?

CHARCOAL AND CORD WOOD

Mining silver ore deep within the Earth, milling it to extract the precious metal, and sending it on its way required energy. The Cortez District ran on heat, produced with charcoal and cord wood cut from the thick stands of pinyon and juniper on the slopes of Mount Tenabo. Pinyon, a hard and resinous wood, was converted to charcoal to generate the massive, sustained heat required for processing ore. Charcoal is basically pure carbon. It burns hotter and produces more heat than an equal volume of cord wood, without the weight and bulk. Meanwhile, juniper fired the boilers for the district's ubiquitous steam powered machinery, along with heating its homes and shelters, baking its bread, and boiling its water.

Steam engines ran practically every machine in the district, including the hoists that brought ore to the surface from deep underground, the air compressors that ran the rock drills, the milling machines that crushed the ore, and the agitators that mixed and stirred the slurry of chemicals and crushed rock. The forty-horsepower engine in the Mill Canyon and the steam-heated pans used in the Washoe process ran from a boiler fifteen feet long and four feet wide. The later Reese River and Russell processes required extraordinary heat to roast each four- or five-ton batch of ore and salt for up to seven hours. Roasting a typical ton of "rebellious" ore required 30 to 60 bushels of charcoal, made from one or two cords of cut pinyon. A modest mill processing 1,000 tons of ore per month would burn anywhere from 30,000 to 60,000 bushels of charcoal during that time.[1]

Mount Tenabo's nineteenth-century pinyon-juniper stands were generally composed of larger, older trees than today's relatively dense, second growth forests. The trees were farther apart, leaving a more open canopy. These forests were one of the district's most important assets, as a May 1864 newspaper article described: "At the foot of Mount Tenabo the Cortez Company owns a most valuable woodland of several hundred acres, which they have had surveyed in accordance with the laws of the Territory, and have had the precaution to further secure it by a substantial fence. This woodland is of easy access and contains wood and timber sufficient to supply the company with mining timber and fuel for steam for many years."[2]

The district's demand for cord wood was equal to its need for charcoal. Crews cut, limbed, and stacked three- and four-foot juniper logs, hauled them by pack mule, wagon, or sled in winter, to stockpiles, or distributed them to boilers, wood-stoves, and cookstoves. An October 1888 Cortez Mines Ltd. ledger entry, for example, listed 7,436 cords of wood among the company's assets. The Tenabo Mill consumed 452 cords of wood that month, while the Garrison Mine used 18 cords.[3]

Making charcoal, also known as "coaling," meant burning the wood very slowly in an oxygen-starved environment. This drove off the moisture, oils, and gases that made up most of its weight and bulk, leaving only charcoal behind. Permanent stone kilns were used for charcoal making in many mining districts, but large earth ovens served this purpose at Cortez. The key to success was to allow just enough oxygen into the kiln or oven to keep the coaling process going, but not so much as to let the wood burst into flames and reduce the entire load to a worthless pile of ash. Ideally, coaling converted the pinyon logs and branches to carbonized, intact versions of themselves—nothing like the charcoal briquettes we envision today. The carbonized wood did invariably break up, but the best product consisted of as many large chunks as possible.

FIGURE 5.1. Typical earth-covered charcoal oven. The location is unknown, however this was a very common method used throughout the country. The tender's camp is in the background. The remains of such camps are commonly found near oven locations. Courtesy of Erika Johnson.

The charcoal makers—sometimes known by the Italian *carbonari*—began the process by clearing and leveling an area up to fifty or sixty feet in diameter. They placed four- or five-foot lengths of cut and trimmed pinyon together on end, forming a large, circular body of upright logs. They then added a second, similar course or laid more wood horizontally over the first. An opening was left in the center of the stack to serve as a chimney. They added a layer of grass and twigs and then covered the entire pile with dirt. The grass and twigs were set fire through the chimney or other openings placed around the perimeter of the oven. This in turn ignited the stacked logs. The charcoal makers continuously adjusted the air flow to the fire by blocking or unblocking the chimney and opening and closing the vents at the bottom of the oven. This maintained a slow, even burn that over the course of several weeks gradually transformed the wood into charcoal. Once they judged the coaling complete, they sealed the chimney and vents to completely cut off the air supply and stop the burning. It took several days to completely extinguish the fire, and workers often dismantled the ovens at night so they could spot and extinguish any live coals. Then they bagged up the charcoal and hauled it away by pack mule or wagon.[4]

The Mount Tenabo woodlands provided basic fuel through the early 1900s. At times, woodcutters and charcoal burners filled the hills for miles around. The need for charcoal ended in 1908, with conversion of the Tenabo Mill to cyanide leaching and the eventual replacement of steam engines with gasoline and diesel power.

FIGURE 5.2. Archaeologists recording a wood pile high on the slopes over-looking Grass Valley. The wood was cut and stacked but for some reason never used. Courtesy of Summit Envirosolutions, Inc.

More than forty years of charcoal making left their mark on the woodland land-scape. The modern forest has regrown, but you cannot walk far without encounter-ing the nineteenth-century pinyon and juniper stumps. The climate has dried them to the core, bleached and cracked them, and eroded the soil from around their roots. But you can still distinguish the individual axe blows that brought the trees down. In other places, you can find equally desiccated stacks of unused wood.

The archaeological remains of charcoal ovens are not as obvious as stumps or woodpiles, but they are not difficult to find. With coaling completed, workers dis-assembled the oven, shoveling away the dirt covering and pulling or raking out the charcoal logs, branches, and other fragments. They left behind an area of mixed soil, black charcoal dust, and fragments of burned and unburned wood. Rain and snow saturated and dissolved the charcoal and stained the soil to an inky blackness, pep-pered with small bits of charcoal. After a century or so, wind and water might bring in clean sediment or even bury the oven, but the dark soil still stands out. There are other clues as well, such as the cleared, leveled area forming the oven platform. On slopes, the charcoal makers often built low retaining walls on the downhill edge of the platform. Shallow ditches, the source of the dirt shoveled over the logs to seal the oven, often ringed the platform.

When we began our archaeological investigation of a charcoal oven, the first step was to open up a shallow trench through the platform. This exposed the con-tents of the oven and revealed structural details like its depth and diameter. A dense mixture of charcoal stained soil, charcoal fragments, and pieces of unburned wood made up the body of the oven. Red soil, oxidized from the heat of the coaling process, marked the lowest layer of the oven.

We studied seventy-eight charcoal ovens in the Cortez District. These were found on the flanks of the mountain and in the thick pinyon stands in gullies and side canyons. They were excavated into the slope, with more than a third of them having some sort of retaining wall on the downhill side to enclose the fill and form the leveled area. The platforms themselves were mostly oval shaped, although some were square, round, or rectangular. The average size was 22 by 20 feet (345 square feet), with the smallest oven 7 by 10 feet (55 square feet) and the largest 60 by 35 feet (1,649 square feet). None of the ovens showed evidence of being used more than once. That is, they lacked more than a single layer of charcoal-rich oxidized soil. The charcoal makers chose to move to a new location for each charcoal firing, rather than setting up a single, large oven site and then bringing in multiple oven-loads of pinyon from the surrounding area.

Our study also took in a broader view of the ovens' distribution across the landscape. They were concentrated in the pinyon-juniper woodlands on the western and southern slopes of Mount Tenabo, but we were also able to identify oven clusters

FIGURE 5.3. An intact layer of burned logs exposed at the bottom of an oven. Courtesy of Summit Envirosolutions, Inc.

within this area. The largest group included sixty-one ovens within an approximately 240-acre area that also encompassed the Garrison Mine and Tenabo Mill. On average, the woodcutters ranged over an area of just under four acres (about three and a half football fields) for each oven.

We were able to determine dates, timing, and sequence for individual ovens covering more than forty years of charcoal production in the district. Researchers during the 1980s had established a tree-ring chronology for pinyon in the Cortez District, which made it possible to determine the age of almost any pinyon log, whether it was in an abandoned woodpile, left unburned in an oven, or used as part of a structure. This method could also be used to date the stumps that were so common across the landscape.

One question we asked involved how the charcoal makers met the energy demands of the district, and what impact this may have had on the landscape. Accounts from the Eureka Mining District, for example, claimed woodcutters and charcoal makers had clear cut an area for thirty-five miles around the town.[5] Would this description apply to Cortez, where pinyon and charcoal had fueled the district for four decades?

Previous research found that the district's charcoal production did include cutting trees over significant portions of the pinyon-juniper woodland, however the result was not the clear-cuts alleged for the Eureka area. Instead, there were episodes of selective cutting, sometimes years apart, as charcoal makers worked through the forest. Tree ring dates from stumps showed that the cutting had been spread out over a number of years. If the areas had been clear-cut at one time, the cutting dates would all have fallen within, at most, a year or two of one another. In addition, living trees that were large enough to have been alive during the mid to late 1800s had been left standing. These trees were selectively passed over even though at the time they could have produced usable charcoal.[6]

Our more recent studies reached the same conclusions. Tree ring dates from the excavated ovens in the cluster mentioned above ranged from approximately 1864 through 1878—a span of about fourteen years. This timing told us the woodcutters were not dealing with an immediate, overwhelming demand for charcoal but instead a lesser but more constant need. They were content to move through an area, building a new oven for each firing, rather than setting up a central, permanent oven or ovens and then bringing loads of wood to it.

These dates made another interesting point. Wood from the large cluster of ovens was cut between 1864 and 1878, and the cluster was close by the Garrison Mine and Tenabo Mill. However, the mill was not built until 1886, which meant the charcoal was destined for other places. The ovens were also near the mouth of Arctic Canyon—the only opening in that part of the Cortez Range, and the route of the pack trail leading over Mount Tenabo to Mill Canyon. In other words, this area produced charcoal for the Mill Canyon mill. The fact that the cutting was spread out over many years also matches the mill's inconsistent demand for charcoal, as its operators dealt with one problem after another. It also meant that later, when the Tenabo Mill did go into operation, the charcoal makers had to widen their range to include the higher elevations and areas to the north and south along the west side of the Cortez Mountains. As time passed, they cut trees farther and farther from the Tenabo Mill. During the period of relatively high production between 1897 and 1904, they expanded into the Dry Hills south of Cortez and the north end of the Shoshone Range.[7]

We also found a written example of how this relatively small-scale production worked. In May 1889, John Rossi contracted with J. C. Latta to provide 15,000 bushels of charcoal for Cortez Mines Ltd. (the English-backed operator of the Tenabo Mill).[8] The length of the contract is not specified, but it states that Latta had a single, six-horse team that he expected Rossi to keep constantly busy transporting charcoal. According to Bancroft, the Tenabo Mill commonly processed about 20 tons of ore per day, or approximately 600 tons per month.[9] We have no exact figures on how much charcoal it took to roast a ton of ore, not to mention other uses, but estimating roughly 30 to 60 bushels would put the Tenabo Mill's consumption at about 18,000 bushels per month. The Rossi-Latta contract, then, could have produced about a month's supply. We have calculated the charcoal ovens' capacity at anywhere from 1,000 to 3,000 bushels, which meant Rossi could have fulfilled his contract with as few as five or as many as fifteen firings, or roughly two to three months of work.[10]

SALT AND LIME

The switch to the Reese River process after the failure of the Washoe method at Mill Canyon created a need for large amounts of salt, which was added during the stamping process and then bonded with the silver when the ore was roasted. The evaporation of ancient Lake Gilbert and the pluvial lakes that formed every year on the playa left naturally occurring salt deposits on the floor of Grass Valley. A *Reese River Reveille* article from November 1864 noted the discovery of a "bed" of salt a few miles south of the Cortez Mining District. It was "found as efflorescence on the surface, similar to the Smoky Valley salt field," and would be "very convenient and valuable for working the mines of Cortez."[11] The newly created Auburn Salt Mining Company laid claim to about 2,000 acres containing the salt field and surrounding ground. The article estimated production of about 10 tons per day—perhaps an exaggeration—but still an indication of the magnitude of the deposit.

We found no documented evidence of salt production from this deposit, however the Tenabo Company did apparently employ Native Americans to gather salt from Grass Valley.[12] Salt was produced by channeling the briny water that collected in the low points of the valley into ponds, where it was left standing until the water evaporated. The salt left behind could then be shoveled into bags or barrels.

FIGURE 5.4. A lime kiln in Cortez Canyon. Summit Envirosolutions, Inc.
Photo by Robert McQueen.

We do not know if these natural sources were fully exploited or if they produced an adequate supply. Salt remained an important element in beneficiation at the
Tenabo Mill, and company ledgers later showed payments to the Desert Salt Company for shipments of salt brought to Cortez via the Beowawe railhead.

The Russell lixiviation process also added lime and sulfur to the list of necessary ingredients. Lime was used in large amounts, along with sulfur, to treat the silver
chloride solution and form silver sulfide. A few steps later, the sulfur was driven off
in a retort furnace, leaving silver bullion behind.

There are no records of sulfur deposits at Cortez, so it had to be bought and
shipped in to the district. Lime was a different story, since it was the major constituent of the limestone that made up most of Mount Tenabo's west slope. Making lime
involved firing limestone at extremely high temperatures (about 1,800 degrees Fahr-
enheit) in specially built kilns. This broke the chemical bond holding the calcium
oxide and carbon dioxide together in the rock. The carbon dioxide escaped as a gas,
leaving the kiln filled with calcium oxide—or finished lime.[13]

At least three stout, cylindrical stone lime kilns were built in the Cortez District. They had thick, buttressed walls and were dug into hillsides for added strength
against the extreme pressure generated by the build-up of carbon dioxide gas. They
were loaded through an opening at the top, from a platform dug into the hillside
above the kiln. A ground-level entryway was fitted with large steel doors. The kilns
were themselves built of limestone, except for the granite doorway arches and the
granite blocks lining the interior.

Workers quarried limestone from nearby outcrops, and at one kiln we found
piles of unfired, angular limestone blocks left behind in an adjacent quarry. At

FIGURE 5.5. Deteriorated barrels and lime residue. Courtesy of Anthropology Research Museum, University of Nevada, Reno.

another kiln, a wagon road linked the slightly more distant quarry to the loading platform. Two of the three kilns were less than a mile from the Tenabo Mill. The third was farther down Cortez Canyon, but alongside the main road leading back to the mill.

For firing, workers loaded the kiln with alternating layers of limestone and cord wood to ensure that flames would completely engulf the rock. The roof opening and the entryway doors were closed, but smaller air vents, with their own steel doors, could be adjusted to control the air and oxygen supply. Spaces were also left within the loaded kiln to provide airflow and allow the release of carbon dioxide. Firing took three to five days of sustained "red heat," after which the kiln cooled and workers shoveled the lime out into barrels for storage or transport directly to the mill.[14]

The adoption of cyanide processing in the early 1900s ended the need for large amounts of lime, although some was still required for the leaching process introduced at the Tenabo Mill in 1908. The Consolidated Cortez cyanide mill also used lime, but for both these operations it was more efficient to buy lime and truck it to Cortez rather than reactivate the thirty-five year old kilns.

WATER

Silver was not the only precious commodity in the Cortez District. Water was just as valuable in the dry, high desert environment. Every prospector, miner, and mill worker needed water to drink, for cooking, and at least occasionally for bathing. And in the age of animal power, the thirsty menagerie of horses and mules could not be ignored. The needs of the mines and mills, in turn, dwarfed anything humans or animals consumed. Every steam engine in the district used water, and the main boilers—such as the Vulcan boiler in the Mill Canyon mill—needed copious amounts to run stamp batteries, agitators, and any number of other smaller machines. Finally, water went into the mix with pulverized ore to make the slurry that flowed of its own accord from one step in the beneficiation process to the next.

At first, Mill Creek provided an obvious solution to the district's water needs. The mines and mill were either right on the creek or close by. As the *Reese River Reveille* reported in May 1864, "The stream running by the place is called Mill Creek and furnishes an abundance of water to supply the steam works sufficient to run one hundred stamps." A network of pipes and tanks distributed water to the mill and the rest of camp. Springs on Mount Tenabo were few and far between, although one, McCurdy's Spring in Crescent Valley, had been claimed and fenced by one of the original locators. The water, according to the same article, "gushes from the mountain side in a fine clear stream of eight or ten inches volume, running a half mile or more and sinking into the sand."[15]

The discoveries on the Nevada Giant side of the mountain, however, complicated matters. There was no water at the St. Louis, the Garrison, or any other west side mine. They depended on mule trains to bring it by the barrel from springs across the valley in the Toiyabe foothills. The only other feasible source was Shoshone Wells, more than a mile and a half west of the mines. But the small community there likely consumed all the water the "wells" produced.

This all changed in 1886 with construction of the Tenabo Mill. It needed a water supply all its own, well beyond anything that could be carried by mules one barrel at a time. Locating a mill closer to the area's springs, or drilling wells and setting the mill up nearby, only meant those same mules would be hauling ore to the water rather than the other way around. The solution was to build the mill close to the mines and find a better way to get water to it.

Two years earlier, with the Tenabo Mill still in the planning stage, newspapers reported that Wenban had begun work on an ambitious project to bring water to the future mill site.[16] The $60,000 system was an ingenious combination of gravity flow, siphon, and pumps that tapped two sources. The first was a spring—appropriately named Wenban Spring—about seven miles southwest of the future mill site. The second source was two artesian wells on the floor of Grass Valley several miles south of the mill. Work on one of the wells began as early as September 1883 when the *Nevada State Journal* reported that Wenban was sinking an artesian well, currently at 550 feet, and expected to encounter flowing water very soon.[17]

Wenban Spring was 110 feet higher in elevation than the Cortez townsite. Water flowed by gravity down from the spring to the valley, and then the siphon effect carried it up the slope to its destination. Water from the wells originated on the valley floor and had to be pumped to the mill. Bancroft noted, "Two artesian wells were sunk in the valley, but the water had to be forced two miles, and raised 417 feet by

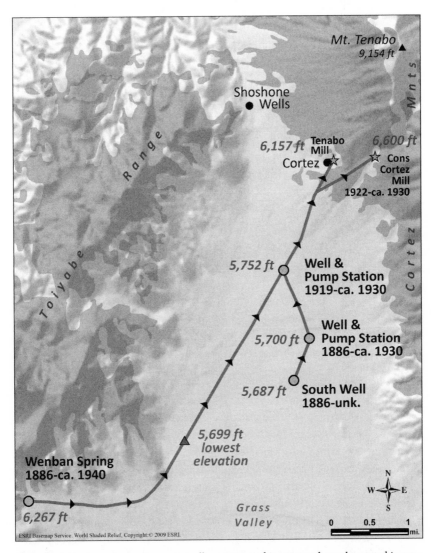

FIGURE 5.6. Cortez water system, originally constructed in 1886 and supplemented in 1919 by Consolidated Cortez. Courtesy of Environmental Systems Research Institute, Inc.

means of a Worthington pump."[18] The steam-powered pump was replaced in the 1920s, when Consolidated Cortez brought in diesel pumps. Imported English glazed steel pipe, seven-eighths inches in diameter, comprised the water lines.[19] The relatively small diameter pipe helped maintain a higher water pressure.

The picture of archaeologists at work often includes trowels, paintbrushes, and even dental picks being used to painstakingly expose artifacts and features. But the sound of heavy machinery is not foreign to our investigations either. We had

FIGURE 5.7. Concrete pump mounts and other features at the main pump station. The upright anchor bolts mark a second pump in the background. Summit Envirosolutions, Inc. Photo by Robert McQueen.

compiled a tentative map of Wenban's water system based on our fieldwork, aerial photographs, and historic maps. Locating actual pipe would confirm our picture of the system. Mechanical excavation with a backhoe gave us a much better chance of success than the much slower and more limited hand excavation.

We dug a series of backhoe trenches perpendicular to what we thought was the pipeline route between Wenban Spring and the valley floor. This uncovered water pipe in three locations, buried anywhere from two and a half to three feet deep. Its two and a quarter inch outside diameter probably conformed to Wenban's English one and seven-eighths inch pipe. A fourth trench came up empty, although this was not entirely unexpected because historic pipe was often recycled once the line was abandoned. In another place, someone had cut the original pipeline and joined it to a plastic pipe for watering modern livestock.

We also investigated the wells and pumping stations that were the heart of the system. These included the south well, the main pump station and well, and the north well. Consolidated Cortez probably added the north well, and a pump, to the existing system in 1919. An inscription on one of the concrete pads at the north pumphouse read "J F Coleman / 1919." According to a Consolidated Cortez report, the company's work did include installing new pumps and tanks for the water system.[20]

Two sets of concrete machinery mounts for the steam engine and pump marked the location of the original pumphouse. Large wooden beams enclosed them and formed the corners of the building's foundation. We found artifacts typical of a workplace setting, such as pieces of belting, fragments of glass pressure gauges, gasket fragments, a large flat file, a hacksaw, and a hand-forged drill bit. Structural artifacts included window glass, nails, and roofing tacks.

We also uncovered the remnants of a firebox for heating the boiler that provided steam power to the pump. It appeared on the surface as a mound of deteriorated firebrick and rock, and a curving alignment of quartzite and limestone cobbles. The mound was actually a brick structure capped and sealed with concrete. Dark gray ash marked the location of the chimney. At least 100 fragments of yellow firebrick and standard orange bricks were found scattered to the west of the firebox. The firebrick was manufactured in England by two different companies, R. Brown and Son (ca. 1836–1938) and Joseph Cowen and Co. (ca. 1823–1904). The orange, common brick was likely made here in Grass Valley. Two long, threaded rods reinforced the firebox and held the smokestack in place.

A collapsed structure just north of the machinery mounts served as a workshop. The remains included a number of flattened cyanide drums reused as siding as well as lumber with wire nails. Other structural elements were discovered as we excavated, including wood posts, large beams, and a row of bricks. The use of recycled cyanide drums as building material, along with the wire nails, indicated this part of the structure postdated the 1908 conversion of the Tenabo Mill to the cyanide process. The buried parts of the structure may have been built earlier and then refurbished with flattened cyanide drums as siding.

A second collapsed, rock-walled room nearby was probably a residence. At least ten courses of irregular, dry laid cobbles comprised the interior walls. It also included a number of work-related artifacts and a few personal items, including fragments of a stoneware Chinese container. There were also twenty-three parts to a cast iron stove.

The archaeological record included evidence of over half a century of operation and the lives of the people who maintained the district's water supply. Our excavation revealed the overall layout of the boiler, steam engine, and pump, as well as the structures housing the machinery and sheltering the workshop and the residents. The various tools and other artifacts, like pipe fittings, hacksaws, sheets of rubber, assorted nuts and bolts, and fragments of the glass covers for instrument gauges, showed the kinds of work that kept the pumps operating.

The main pump station and well were key elements of the system from the Tenabo Mill's first day in 1886 until 1915 when the mill closed. In 1922, Consolidated Cortez modified the pipeline and redirected water to their new cyanide mill and used the system until they shut down in the early 1930s. A 1938 US Geological Survey topographic map of Cortez still labeled the site as a "pump station."[21] The map makers did not know the pumps had been stilled years before, but we fortunately had more accurate information, which turned out to be a unique mix of both archaeology and memory—as we will explore later.

Brick buildings represented an important step in the evolution of any mining district. As brick construction replaced tents, dugouts, and log and rough board buildings, the camp took on an air of maturity and permanence. Bricks demonstrated a serious financial commitment and showed faith in the future. Brick banks, hotels, and courthouses meant people expected to be on the scene for years to come, successfully carrying on with their business. Brick industrial construction likewise showed confidence and faith in the district's mining potential. In the mills, thick brick walls set on foundations of dressed stone blocks withstood the heat and vibration of the roasters and crushers. Brick was well suited for forges, assay furnaces, repair shops, explosive storage, and the towering smokestacks that carried acrid and dangerous fumes away from a mill and its surroundings.

Unlike some of Nevada's more established mining district towns, like Austin, Eureka, and Virginia City, Cortez never developed a "Main Street" or significant "downtown." Brick construction was limited to the Tenabo Mill and some ancillary structures, along with a few chimneys in the townsite. The Mill Canyon mill was reportedly made of stone, most certainly native rock gathered from the canyon itself. Other structures in Mill Canyon were wooden, including some with palisade-style walls of vertical pinyon and juniper logs. Dugouts, stone buildings, and frame houses were the most common structures in Shoshone Wells.[22] Our partial survey of the Cortez townsite showed two dozen dugouts, a handful of standing wood frame and adobe structures, and traces of other residences from foundation outlines to scatters of decaying lumber.[23]

Wenban wanted the Tenabo Mill to be an impressive, state-of-the-art operation. Brick, along with being well suited for mill construction, also presented a confident and sophisticated appearance that wood or native rock lacked. Brick was cheap and easily available, providing the Cortez landscape included concentrations of sticky, fine-grained sediment within reasonable distance of the mill site. With that, the only further requirements were water, fuel for firing the bricks, and brickmakers.

Lloyd High, the last resident of Cortez, said in a magazine interview that the self-reliant folks of the district made their own brick from a clay deposit about seven miles southwest of Cortez. On another occasion he stated they made brick for the mill by hand, in molds, near Wenban Spring.[24] But the location of the Cortez District's brickworks remained forgotten and undiscovered until the archaeological investigations of the early 2000s.

Brickmaking definitely left its mark on the landscape. Clay soil was excavated, mixed with water, molded into bricks, and then fired in mass quantities. But the remains of a one-hundred-plus-year-old brickworks would not necessarily be easy to find in the thick grass and sagebrush covering Grass Valley's lower elevations. It might also be difficult to distinguish a brickworks from the ruins of collapsed or burned brick buildings. Nevertheless, during surveys in 2003, archaeologists discovered two potential brickmaking sites—one on the alluvial fan at the west edge of Grass Valley and one about a mile away on the valley floor.

The alluvial fan site included a number of cut-and-fill earth platforms and a depression about 12 feet on a side and a few inches deep. The other location had a number of low, eroded mounds that turned out to be piles of deteriorated brick. Small, dissolving fragments were scattered across both sites, leaving a distinct reddish hue to the soil. We only found one complete brick among the hundreds of fragments. The site also included a large, kidney-shaped clay pit—100 feet long, 50 feet wide, and 5 feet deep.

Research into nineteenth-century brickmaking guided our further exploration and excavation of these sites.[25] Brickmaking began at a source of naturally occurring clay, which was dug up, mixed with water and temper, packed into molds, dried, and then fired. A number of different substances worked as temper to keep the brick from cracking during firing. These included sand, charcoal, lime, ash, chalk or even coal dust. The clay was mixed in a pug mill—basically a wooden tub with a rotating central shaft with several mixing blades attached. Workers turned the mill by hand or used a horse harnessed to an armature that rotated the shaft as the horse walked in a circle. A quantity of clay soil could also be piled on the ground in a "soak heap" and mixed with water and temper by hand, using a spade to turn the pile or employing a three-foot-long iron bar like a sword to repeatedly "cut" the pile.

Workers packed the clay mixture into rectangular, open-faced molds. Excess clay was scraped off, or "struck," from the exposed face, and the molded bricks were laid out to dry. Once dried, the "green" bricks were ready to be stacked in a clamp for firing. A brickmaking clamp is essentially a self-contained, open air oven. The bricks are stacked with spaces between them for heat circulation during firing. Empty chambers honeycombed the interior of the clamp and were packed with fuel—at Cortez it was juniper cord wood. A layer of previously fired brick sealed the exterior. As the fuel burned, the various chambers and spaces distributed the heat throughout the clamp, partially vitrifying and hardening the green brick. The process took several days from ignition to final cooling.

Brick clamps were usually quite large, at least 10 to 12 feet high and 30 or more feet square. They produced tens of thousands of standard size (8 by 4 by 2½ inch) bricks at each firing. The mining camp version of the process was uneven at best, leaving behind a fair share of broken or underfired bricks. Underfired bricks could be loaded into the next clamp or collected and refired in smaller clamps. The discarded, disintegrating brick fragments and the red stain they left with the soil became the distinctive clues allowing us to identify brick-making sites.

FIGURE 5.8. Brickmaking in Colorado, with a horse-powered pug mill in the foreground, rows of drying bricks, and two large brick clamps in the background. Courtesy of Salida Regional Library, Centennial Photo Archive.

We excavated the buried remains of three brick clamps. At one, we discovered a large, intact portion of the floor, essentially the clamp's bottom layer of brick, which had been left in place for the next firing. The heat and mixing of charcoal and unburned wood fragments had blackened several inches of sediment beneath the floor. We determined the clamp was about 27 feet square, with at least seven flue tunnels on the floor level. The flues were clearly distinguished by ashy remains of the burned fuel, as they channeled air and fire throughout the clamp.

The second brick site further south in Grass Valley only included a portion of its original clamp floor, although rows of fired brick formed a partial outline around it. An area of blackened and burned soil about 27 by 21 feet marked the body of the clamp. The mounds of decaying, disintegrating brick that were first discovered at the site were piles of ash and discarded brick from the main clamp.

These two brick-making sites supplied the Cortez District, including brick for building the Tenabo Mill. The two larger clamps we investigated were each capable of producing about 35,000 to 40,000 bricks per firing, with the smaller clamp producing 15,000 to 20,000 bricks. The Tenabo Mill probably required at least 100,000 bricks in its construction, and other needs throughout the district added at most another few thousand. The brickmakers could have met this need with a relative handful of firings, although each one required intense, concentrated effort mixing clay, molding bricks, stacking the clamps, and unloading and transporting the

FIGURE 5.9. Exposed brick clamp floor. The light-colored ashy bricks mark the flues. Summit Envirosolutions, Inc. Photo by Robert McQueen.

FIGURE 5.10. Excavating the clamp floor. Summit Envirosolutions, Inc. Photo by Robert McQueen.

finished product. We did not find evidence of long-term use of the clamps or signs of substantial, extended occupation of the sites.[26]

Artifacts found with the clamps included tools and a few cans, bottles, and other domestic items. The most unique tool was a 14-foot-long poker, made by welding a hand-forged point onto a length of narrow metal tubing, which was also bent into a loop to form the handle at the opposite end. At 14 feet long, it could have reached any point to stir the fuel from one side or the other of the 27-foot-long flues.

Demand continued for brick after construction of the Tenabo Mill. Company ledgers showed that in twelve years from 1889 to 1901, 76,422 bricks were bought for use in the Cortez District, including 72,700 at the mill, 2,700 at the mine, and 1,000 in the construction of other buildings. During the same time, the Tenabo Mill and Mining Company received an additional 132,000 bricks, in seven separate deliveries, some spaced several years apart. The last major brick purchase noted in the Cortez company ledger was for 40,800 bricks delivered during the first three months of 1900.[27]

Brick construction contributed to the efficient design and operation of the Tenabo Mill, and it played a critical role in one of the most successful periods of the Cortez District's history. Brick gave the building a substance and permanence that could not help but impress world-class investors such as Bewick-Moering. The 1908 conversion of the Tenabo Mill to the cyanide leaching process did not require any additional brickwork, and by the 1920s brickmaking in the district had long been abandoned. Consolidated Cortez built their mill of corrugated sheet metal, supported by a wood superstructure and concrete foundation. Cheap, quick sheet metal construction turned handmade bricks into an unaffordable luxury.

TRANSPORTATION

If you walked or rode horseback in the Cortez District, maybe going to work, making some small delivery, or returning home with supplies, you seldom had to share the road. On occasion, you might have to step aside for a string of pack mules—or be pushed off the trail as they were famous for doing. Or you might give way to freighters, their teams pulling multiple wagons piled high with goods of all kinds, or wagons heaped with bags of charcoal.

As busy as the district was, Mount Tenabo was spacious enough to comfortably envelope the prospectors, miners, and woodcutters, and all their endeavors. The haphazard spiderweb of roads and trails connecting it all sprang up piecemeal, whenever a new find or new resource needed to be moved from one place to another. They naturally trended downhill, merging as they approached the Tenabo Mill or the Cortez townsite or coalescing to funnel traffic to Shoshone Wells, out Cortez Canyon, or down Grass Valley toward Austin.

Moving things around within the district was half the transportation story. The other half linked Cortez to the outside world to meet needs as mundane as a week-old newspaper or as vital as the recruitment of skilled tradesmen. Most everything people ate, used, or simply desired had to come through Austin or Beowawe. And more important, even the shiniest silver ingot was worthless until it reached the outside world where it could be turned into money.

When Andrew Veatch and his men set out from Austin in 1863, they followed the stage road east a few miles toward the Simpson Park Mountains, then turned north into Grass Valley. The stage road was part of the Overland Trail, also known as the Central Route, which was one of the two cross-country routes through the Great Basin. The other was farther north along the Humboldt River. Captain James Simpson surveyed a wagon road along the Central Route in 1859, and it quickly developed into a major travel corridor between Salt Lake City and Carson City. It became known as the Overland Trail, after the Overland Mail Company, which ran passenger and mail service along it.

The section of the Overland Trail between the Comstock and Carson City area and Austin was also known as the Reese River Road. In addition to the passenger, mail, and freight service the Overland Stage provided, the Reese River Road was also a major freighting route. It ultimately linked the mining centers of Austin, Eureka, and other areas in central Nevada to California. As the initial strikes in these districts developed into more substantial mining and milling endeavors, the demand for supplies and equipment turned the Reese River Road into one of the busiest wagon roads in the West. In 1865 the Overland Mail Company "carried between Virginia City and Austin 5,840 passengers" while other freighters accounted for 7,620 tons of merchandise, machinery, and lumber during this same year.[28]

Simeon Wenban and the original principals of the Cortez Gold and Silver Mining Company shipped the components of the Mill Canyon mill from San Francisco and Sacramento to Austin on the Reese River Road. Teams completed delivery of the sixty tons of machinery to the mill site by April 1864, using what must have been a rudimentary wagon road north from Austin through Grass Valley. The *Reese River Reveille* had lamented Cortez's remoteness and the lack of good roads in July 1863, noting the consequent difficulty getting mining supplies delivered to the district.[29] But by the spring of 1864 the same newspaper noted wagon freighters and travelers now easily traversed the improved roads to Cortez.[30] Five years later, in 1869, the General Land Office maps of the Cortez District showed an extensive road network in Mill Canyon, lower Crescent Valley, Grass Valley and the western slopes of Mount Tenabo.

Transporting ore from mine to mill was a major challenge during the Cortez District's first two decades. Pack mules took the first Mill Canyon ore seventy-five miles south to Austin for milling, and this remained an option for years to come due to the recurrent problems at the Mill Canyon mill. Wenban delivered his break-through ore from the St. Louis mine to Austin by mule train in 1867. Even with a functioning mill and successful production from the Garrison and other west side mines, ore still had to be packed along trails over the south shoulder of Mount Tenabo and down into Mill Canyon for milling.

Pack mules provided an exceptionally versatile form of transportation. They were especially well suited for reaching remote locations and crossing difficult ter-rain. And long strings of mules could carry a surprising amount of ore, with Cortez pack trains delivering fifteen or twenty tons with each trip. Mules were important to the early development of a mining district, because they could make due with trails instead of engineered wagon roads, and they required less work and operating cap-ital than teams of draft animals and ore wagons. At Cortez and elsewhere, wagon roads replaced mule trails and freighters and teams took over the transportation business as the need developed for heavy machinery and large volumes of supplies and equipment.

WATER AND CHARCOAL TRAILS

We discovered a number of mule trails that showed the utility of pack mules for accessing difficult locations and hauling cargo that did not lend itself to bulk trans-port. One trail wound its way from a spring in the Toiyabe Range to the settlement of Shoshone Wells. It was likely used by pack trains laden with water barrels and bound for Shoshone Wells or the Garrison Mine.

Another trail passed east from Cortez through a heavily wooded area on the mountain's alluvial fan and ended near several charcoal platforms. It was a simple, meandering scar on the landscape traveled by mules loaded with cord wood and charcoal on their way to the Tenabo Mill and the west side mines. The evidence of its use included mule shoes and artifacts marking a small, spring-side campsite where muleskinners rested and watered their animals.

Other forms of commercial transportation developed as Mill Canyon, Sho-shone Wells, and then the town of Cortez became home to a population of miners, mill workers, their families, and others involved in various supporting occupations. When Simeon Wenban moved his family to Mill Canyon in 1864, there was no com-mercial transportation between Austin and the district. But by July 1867, George Rus-sell had begun stage service to the Cortez District from Austin.[31] The stage departed Austin on Monday and returned on Saturday, often with a load of bullion. That schedule tells us the trip took about a day and a half, with an overnight stop at one of the ranches along the way. Service increased to twice weekly later in the year and also included mail delivery. Newspapers of the day often published lists of arriving and departing passengers as well as various wagon freighters and their cargoes. The

AUSTIN and CORTEZ

STAGE LINE!

Semi-Weekly Line of Stages, carrying Passengers and Freight

From and to Austin and Cortez. Will leave Austin on Wednesday and Saturday, and Cortez on Monday and Friday of each week.

OFFICE—At Horton & Sawtelle's store.

GEORGE RUSSELL.

Office in Cortez at Russell & Brother's store.

Austin, Sept. 10th, 1867.—tf

FIGURE 5.11. Advertisement for George Russell's Austin and Cortez Stage Line. Courtesy of Nevada Historical Society

Reese River Reveille frequently noted Simeon Wenban's travels as he went back and forth to San Francisco on business.

Transportation to Cortez and all across Nevada changed dramatically in 1869. The Transcontinental Railroad was completed that year, and it streamlined east-west travel across the state. It also changed travel to interior locations like the Cortez District. The railroad did not by any means eliminate the need for freight wagons, which remained the mainstays of transportation well into the twentieth century. But it seriously diminished the importance of the long-haul, east-west wagon routes like the Reese River Road. It became cheaper and more efficient to bring equipment and supplies into the state by rail, from either the east or west, and deliver it to railheads like Beowawe. It was then off-loaded to freighters who took it to the final destination north or south of the railroad. For Cortez, the short thirty-five-mile trip through Crescent Valley from Beowawe replaced the arduous wagon haul over numerous mountain ranges, including the Sierra Nevada, from California to Austin, and the subsequent seventy-five-mile journey north to the district.

By the 1880s, significantly more traffic used the Crescent Valley wagon road from Beowawe than the Grass Valley road to Austin. Ranchers in the Beowawe area found a ready market for their hay and grain to feed the teams pulling the freight wagons loaded with lumber and merchandise bound for Cortez.[32] Because of the railroad, supplies, mail, news, and people all arrived faster than ever before. The railroad also made it possible for Cortez mines to contract with smelters and mills in Salt Lake City and northwest Utah to process their ore. A Reno newspaper article in 1888 announced the construction of an enormous "prairie schooner" designed for hauling quartz between Cortez and Beowawe.[33] More details were found in a later article, which described a "huge land schooner" capable of hauling 18,000 pounds. "The hind wheels stood over seven feet high, the spokes were clubs, the tires almost as wide as a dinner plate." The lead mules on the team that pulled the wagon were Kentucky mules, one standing "fully six feet five inches at the shoulders."[34]

In good weather, freight and people disembarking at Beowawe could expect a day's wagon trip to Cortez, while large, cumbersome shipments took maybe three or four days. In the winter it might take several days, as snow blocked the roads and then turned them into quagmires when it melted. Correspondence between the store at Cortez and Beowawe showed how unreliable winter transportation could be. A letter from the winter of 1913 noted, "This will be Minoletti's last trip for some time until the roads get better. There is so much snow here now that we won't be able to get anything in for some time after this trip." The store clerk repeated the request: "We won't be able to get anything in here after this trip so please send all you can. Send as much ham and bacon as you can spare."[35] The order also requested coffee (whole [bean], or canned if not available), hay, canned milk, canned fruit, canned tomatoes, and cigarette papers.

In the early twentieth century, gasoline or steam-powered vehicles began replacing animal power. An August 1907 *Reese River Reveille* article announced the first trip of a "traction engine" between Beowawe and the Tenabo mining camp, a newer mining town in the Shoshone Range about fifteen miles from Cortez.[36] The engine brought merchandise and equipment south and then returned north pulling a string of ore wagons. A subsequent article noted the engine's departure from Tenabo to the railroad loaded with forty tons of ore, accompanied by four teams, each hauling twenty to thirty tons of ore.[37] A mine report in 1910 noted that a five-ton auto truck was being used to make regular shipments from the Cortez Mine to Beowawe, where the ore was sent by rail to smelters in Salt Lake City.[38]

People have traveled in and around the Cortez District since the early 1860s, and they continue today in all varieties of on- and off-road vehicles. They have left innumerable "roads" across the landscape, everything from faint tracks through the brush to well-used two-tracks to graded, maintained county roads. An archaeological view of transportation in a place like Cortez involves classifying these transportation features according to their role within the larger transportation network. We divided the historic roads in the district into three groups: major wagon roads linking the district to the outside world; intermediate roads connecting different places and activities within the district; and the trail network, or resource roads, which brought specific, otherwise inaccessible or difficult-to-reach locations into the overall system.

Distinguishing historic roads, or identifying the historic use of a modern road, required clues, often in the form of roadside trash, and some detective work with old maps and records. Thankfully for archaeological study, travelers never thought twice about simply throwing away empty bottles and cans until well into the twentieth century, when automobile travel became the norm, along with campaigns to stigmatize and eliminate littering. Historically, horse, wagon, and stage travel left a trail of wagon hardware, horse, mule, and oxen shoes. Later came automobile parts, which added to the background of roadside debris and campsite litter.

OLD STATE ROUTE 21

This was the first wagon road linking Austin and the Cortez District. It approached from the south through Grass Valley along the east edge of the valley floor, aiming at the head of Cortez Canyon. It passed beneath Mount Tenabo, and short access roads linked it to the Cortez townsite and Shoshone Wells. It went through Cortez Canyon to Crescent Valley, where it continued north to Beowawe. A branch to Mill Canyon skirted the south end of the valley and then ascended Mill Canyon to the original camp and millsite.

We identified much of this road by recording it on the ground and then comparing its position with a series of historic maps to confirm that it was in the same place as the original road through the valley. The historic maps included the General Land Office (GLO) map from 1870, a second GLO map from 1924, and a 1938 US Geological Survey topographic map. The road also appeared on a 1954 aerial photograph, taken shortly after the modern county road had been graded parallel to the historic road.

The road segment appeared as a well-worn roadway, relatively clear of brush and eroded slightly below the surrounding ground surface. Artifacts scattered along it included tin cans and bottle fragments dating as early as the 1880s, along with horseshoes, wagon hardware, and automobile parts. The road was part of the original link between Cortez and both Beowawe and Austin. The connection to Beowawe became more important with completion of the railroad, but travelers still went back and forth to Austin, which was the Lander County Seat until 1979. The road was designated State Route 21 with the advent of automobile travel and the establishment of a state highway system in the late 1910s and 1920s. State Route 21 linked the Lincoln Highway (the old Central or Overland Route, later US 50) on the south to the Victory Highway (US Highway 40 and later Interstate 80) to the north. As a state highway, the road received better maintenance, but it was never paved. It was decommissioned in the 1940s and abandoned by 1953, which reflected the diminished importance of Cortez during that period.

CORTEZ ROADS

Two short two-track access roads connected Cortez with the main Grass Valley wagon road. One headed northwest from Cortez toward Shoshone Wells and the other went southwest until it, too, intersected the main road. The first was only a mile long, but it was the initial leg of the connection from Cortez to Beowawe, and the rest

FIGURE 5.12. Old State Route 21. Courtesy of Summit Envirosolutions, Inc.

FIGURE 5.13. This key fob from Room 8 at the Cortez Hotel
was found in one of several dumps along the road out of town.
Courtesy of Summit Envirosolutions, Inc.

of the world. When we first identified the road we documented up to five distinct, parallel sets of tracks as well as several roadside dumps. The parallel routes, some resembling shallow ditches, represented a unique phase in the evolution of undeveloped wagon roads. Repeated use churned axle-deep, dust-filled ruts into the soft soil, which rain or snow turned to tenacious, unforgiving mud. The ruts also caught and channeled runoff, which eroded them even further below the surrounding ground surface. The solution was to simply abandon the tracks when they became impassable and move the road a few yards to one side or the other. The process was repeated as many times as necessary, resulting in multiple parallel routes.

The other access road led southwest to State Route 21, and carried traffic from Cortez to Austin and then to points east and west. Grass Valley ranchers bringing hay, meat, and produce to Cortez would have used this road.

NOTES

1. Hattori et al. (1984, 5–6).

2. *Reese River Reveille*, May 5, 1864, 1.

3. Johnson and McQueen (2016, chapter 96: Charcoal and Cord Wood Production, 9; citing Cortez Mines Ltd. ledger).

4. Reno (1994, 1996).

5. Hattori and Thompson (1987, 65, citing Lanner 1981, 124–130); Young and Budy (1979, 117) are among those who believe pinyon-juniper woodlands were clear-cut in a sixty-mile radius around Eureka.

6. Hattori et al. (1984, 28).

7. Johnson and McQueen (2016, chapter 96: Charcoal and Cordwood Production).

8. Contract between John Rossi and J. C. Latta, Cortez Nevada, May 14, 1889.

9. Bancroft (1889, 253).

10. Johnson and McQueen (2016, chapter 96: Charcoal and Cordwood Production, table 4: Summary of Charcoal Platform Data).

11. *Daily Reese River Reveille*, November 15, 1864, 1.

12. High (n.d., 12).

13. Perry et al. (2007); Johnson (2008).

14. Perry et al. (2007).

15. *Reese River Reveille*, May 7, 1864.

16. *Daily Nevada State Journal*, June 11, 1884, noted Wenban was building new mill and laying pipe; Bancroft (1889, 253) gave the cost as $60,000.

17. *Daily Nevada State Journal*, September 21, 1883.

18. Bancroft (1889, 253).

19. Bancroft (1889); High (n.d., 7).

20. Consolidated Cortez Silver Mining Company (1923).

21. United States Department of the Interior Geological Survey (1938, reprinted 1970); Nevada Cortez Quadrangle 15-Minute Series.

22. Hardesty (1988, 84–86).

23. Johnson and McQueen (2016, chapter 103: Residential Features).

24. Murbarger (1959).

25. Gurcke (1987); Garvin (1994).

26. Johnson and McQueen (2016, chapter 98: Brickmaking).

27. Johnson and McQueen (2016, chapter 98: Brickmaking).

28. Angel (1881, 467).

29. *Reese River Reveille*, July 8, 1863.

30. *Reese River Reveille*, April 7, 1864.

31. *Reese River Reveille*, June 10, 1867.

32. *Nevada State Journal*, January 10, 1891.

33. *Nevada State Journal*, April 18, 1888.

34. *Elko Free Press*, September 29, 1888.

35. Johnson and McQueen (2016: chapter 105: Consumer Behavior and Material Culture Practices in the Cortez Mining District, quoting an exchange of letters between Weber Company, Beowawe Mercantile Company, and Tenabo Mill and Mining Company, January 30, 1913).

36. *Reese River Reveille*, August 3, 1907, 2.

37. *Reese River Reveille*, September 14, 1907, 3.

38. Emmons (1910); Parker (n.d.). Parker's "Report on the Cortez Mine" was probably written during the preliminary phases of Consolidated Cortez' development. He notes ore and supplies are hauled to and from Beowawe in five-ton trucks.

Chapter 6

Many Lives, Many Stories

People leave their mark on history in a number of ways. Successful mining men like Simeon Wenban found their skill at reading geology and ferreting out precious metal and their business acumen praised at every turn by mining journals, newspapers, and magazines. The society pages reported dutifully on their social lives and chronicled their family's every birth, death, and marriage. Others, like Andrew Veatch, became legends in the mining camps, their success giving hope and meaning to men searching the bare, sun-scorched rock for their own fortunes. Some managed to distinguish themselves in ways that, for good or bad, got them in the history books. But these are the rare exception to the countless people whose lives took them to Cortez, and are now invisible or known only from an entry on a census schedule, a signature on a contract, or a brief mention in a newspaper article. They were the ones who labored all across the rocky mountainside with picks, shovels, and black powder, hoping to become rich like Simeon Wenban. Or they went down every day into the dark heart of Mount Tenabo and, in return for a daily wage, drilled and blasted the silver-bearing rock and sent it to the surface. They chopped at trees and dragged them to woodpiles; wrestled charred logs from collapsed, burned-out charcoal ovens; breathed the hot, acrid dust from a lime kiln; bent over a brick mold; hammered an anvil; cursed a team of mules; labored at home, minding children, cooking dinner, and mending clothes; or were themselves children lost in some imagined game.

Where do we look for their names and stories? One place, as archaeologists know, is right under our feet. Archaeology uncovers the story of everyday life, and those otherwise anonymous people only needed to drop something on the ground, toss aside an empty bottle, set down a tool, or scrape together some kind of shelter to be included. It then becomes our job to put the evidence together and tell their story, even though, like a broken jar or plate, some pieces will always be missing.

WHO ARE THE PEOPLE?

The population of Mill Canyon, Shoshone Wells, the town of Cortez, and the west slope of Mount Tenabo never numbered more than a few hundred. But the people living in the Cortez District were typical of mining camps across Nevada, large and small. Single men, accompanied rarely by women and even fewer children, made up the bulk of the earliest residents. As the effort shifted from prospecting and exploration to more dependable work in the mines and mills, true communities began replacing the original camps. Populations, and their occupations, grew and

FIGURE 6.1. A reassembled crock, with pieces missing, and a fragmentary
remnant of a decorated plate. Courtesy of Summit Envirosolutions, Inc.

diversified. The ethnic composition of the Cortez District was also typical for the
time. One newspaper article described the first locators as exceptional men who represented, in their nationalities, "the American, English, Irish, German and Swedish
peoples."[1] The population soon included other Americans and Europeans, Italians,
Chinese, Mexicans, and the original Native American inhabitants.

Mining camp life tended to magnify ethnicity and nationality because everyone,
with the exception of the Native Americans, was a newcomer and many were complete strangers to rugged life on the isolated mining frontier. The California Gold
Rush had drawn thousands of people from not only the eastern United States but
Europe, South America, Asia, and Australia. When this same wave of fortune hunters
and assorted colleagues from other walks of life descended on the Comstock and
then central Nevada, they competed with one another for land, mining claims, jobs,
and social status. Newspaper writers and chroniclers of the period were never shy
about pointing out and expressing opinions on the ethnic groups who found themselves thrown together in places like the Cortez District. Within the restrictions and
prejudices of the time, ethnicity was the most obvious, effective way to identify yourself and others, and it often served as the common bond that held otherwise dispa-
rate individuals together.

Ethnicity is also a valuable archaeological and historical tool for identifying how
specific groups of people both influenced and were influenced by the history going
on around them. People can leave their own ethnic "signature" in the archaeological

record. The Chinese, for example, preferred their porcelain rice bowls and uniquely shaped soup spoons to Western ceramics or metal utensils. Italians made bread in outdoor stone ovens just as they had in the old country. Native Americans continued using wooden poles to knock pinyon cones loose and grinding stones to prepare pine nuts.

Chinese

The Chinese had a unique and important role in the Cortez District. They made up the core of the labor force during the district's most productive years, from the 1870s through the early 1900s. Few if any of Nevada's nineteenth-century mining districts lacked Chinese participation in one form or another, although Euro-American prejudice and economic self-interest restricted the Chinese to less lucrative, menial jobs and service occupations.[2] But Cortez was different.

Simeon Wenban employed Cornish miners during the initial development of the Cortez District, but sometime between 1870 and 1873 he replaced them with Chinese workers. Cornwall, the Cornish homeland, is an English county at the southwestern tip of Great Britain with a long tradition of underground tin and copper mining. Cornish immigrants brought their hardrock mining experience with them to the New World. For many years, the Cornish were prominent in northern California and Nevada and were renowned for their mining skills. But Wenban's Cornish became "turbulent and riotous" when his expenses left him unable to meet his payroll. The Chinese replacements not only worked for lower wages, but they assisted him with his temporary financial embarrassment "by waiting for a portion of their pay." Thus began an enduring relationship between the Wenban and the Chinese who, despite his stated objection to "the importation of Asiatic labor," became Wenban's acknowledged friends and the backbone of his workforce.[3]

Wenban's employment of Chinese as underground miners went counter to widespread racial discrimination and opposition from Euro-American miners who used their union organizations, threats, and intimidation to keep the Chinese from competing for relatively high-paying mining jobs. The Chinese were mostly restricted to placer mining and were often only allowed in areas whites had previously worked and abandoned. As districts industrialized and began developing urban centers, the Chinese were relegated to service occupations like cooking, laundering, housekeeping, and menial labor. As laborers-for-hire, contractors, or merchants, Chinese could only work in markets their non-Chinese competitors were either unable or unwilling to serve. Anyone violating this code was subject to censure, if not outright threats. An 1881 Reno newspaper noted Simeon Wenban's employment of a number of Chinese to work at his mines and stated if he did not discharge them "steps will be taken to compel them to go."[4] Wenban ignored the warning and kept Chinese on the payroll for another twenty years. As late as 1896, the *Reno Evening Gazette* complained, "Sixty Chinese from Carson have gone to Cortez to work in the mines," potentially taking jobs of a hundred "white men."[5]

The 1870 US Census counted only one Chinese resident in the small Mill Canyon community, a cook named "A Sen." But by the 1870s or early 1880s, and possibly as early as 1872, Chinese had taken up residence in Shoshone Wells and inhabited a cluster of rock dugouts alongside the Garrison Mine. In 1881, they were prominent enough in the Cortez District to attract the unwanted attention of the area's newspapers. That same year, Thompson and West's *History of Nevada* noted that in Cortez "Most of the labor in the mines is performed by Chinamen."[6]

Lander County property records show that by 1885 Chinese had established themselves in the Shoshone Wells community. Tax rolls list Chinese ownership of cabins, houses, and at least one store in Shoshone Wells. Unfortunately, the 1880 Census for Cortez has been lost, and the complete US Census of 1890 was destroyed through accident and negligence in the 1920s.[7] But Tenabo Mill and Mining Company payroll records documented Chinese work for Wenban's company during the late 1890s. The rolls typically listed about forty workers, most hired through a Chinese labor contractor. Three-fourths of them worked in the mines and the remainder in the mill.

The US Census of 1900 offered the most detailed look at the Cortez District's Chinese population, although the Chinese residents of Shoshone Wells were either overlooked or left out. In 1900 there were 83 Chinese residents, in 20 households, out of a total population of 274.[8] Half the male Chinese worked as mine laborers and about one-quarter as mill laborers. Others engaged in a cross-section of service occupations and small businesses, including cooks, barbers, clerks, butchers, and storekeepers. There were only six Chinese women and no children. The women's occupations were laundress, seamstress, housewife, and housekeeper.

Chinese employment and wages were recorded on payroll ledgers from the Tenabo Mill and Mining Company during the late 1890s. A contractor operating as the Sing Kong Woo Company supplied about forty-three laborers per month, with thirty of them working in the mine and thirteen in the mill. They worked six days a week for $1.50 to $1.75 per day. One turn of the century Cortez resident noted that around 1905 a Chinese man worked the important water pumping machinery in Grass Valley.[9]

There was a sudden exodus of Chinese from the Cortez District in 1904, probably the result of a drastic reduction in the overall work force at the Tenabo Mill and Mining Company. Records show that by June 1904 there were only two Chinese workers on the payroll, and in July, with the total workforce reduced to six, only a single Chinese remained. This minimal workforce continued through the last payroll entry in April 1908. Converting the mill to cyanide leaching that same year did not include opportunities for Chinese workers. By the 1920 US Census, there was, as in 1870, one Chinese resident in the district and he was a cook.

The loss of jobs at the Tenabo Mill and Mining Company was decisive in the Chinese departure from Cortez, but other, long-term conditions also played a role. Foremost among them were the exclusion laws, first passed in 1882 and renewed every ten years until they became permanent in the 1920s. They ended Chinese immigration, and not only kept additional workers out of the country but also women, children, or any other family members of resident workers. In addition to this steady attrition, most immigrant Chinese considered themselves "sojourners" and never intended to live permanently in America. Their objective was to make a living and hopefully acquire enough savings beyond that to return to China and comfortably live out their lives. The exclusion laws prevented the replacement of those sojourners who did achieve their goals and return home. At Cortez, once employment with Wenban's company ended, the district had nothing to offer the Chinese.

Prejudice and segregation restricted the Chinese to certain neighborhoods, or "Chinatowns," throughout the West, but they also remained somewhat apart from the larger community by choice. The non-Chinese population interacted with them when mutually beneficial business or economic opportunities arose. But the Chinese maintained their culture, the language they spoke, the foods they ate, the objects they owned and used, their dress and appearance, and even their medicines and drugs. They continued their own mortuary practices and cemeteries almost everywhere they lived, including Cortez.

The Chinese adherence to their own values and culture left a clear signature in the archaeological remains of their private lives and day-to-day activities. For example, a traditional diet including rice, fish, soy sauce, and tea, came with rice bowls, porcelain soup spoons, stoneware soy sauce jars, and tea cups. These artifacts were all manufactured in China, along with other personal effects like clothing, beads, or coins. The Chinese practice of smoking opium also left behind a distinctive set of artifacts, including opium tins, pipes, and other paraphernalia.

The Chinese did not maintain their traditions in complete isolation. One of our main research questions was how much Chinese life at Cortez reflected traditional Chinese culture and how much it reflected aspects of Western culture they had adopted. Archaeological studies throughout the West have shown that foodways, what people ate and drank and the things they used in storing and consuming their food and drink, were an excellent measure of acculturation. Food has long been a vital part of ethnic identity, and it is one of the last things immigrants give up in their new environment. Fortunately for us, food containers, pots and pans, and tableware are often among the best preserved artifacts in historic sites.

The ledgers of the Cortez company store also provided insights into Chinese food consumption. The accounts covered the period between June 1885 and April

1891, and they listed buyers and their purchases by name. The Sing Kong Woo Co., one of the main Chinese labor contractors, appeared frequently, indicating the company shopped for their employees and then undoubtedly made a corresponding deduction from their pay. The Chinese mostly bought beef, pork, rice, liquor, tobacco, and soap. Pork was the primary meat staple in nineteenth-century China, but beef was cheaper here, so the turn to beef might have been just as much expediency as a cultural adaptation. The Sing Kong Woo Co. was the primary Chinese purchaser of liquor, notably whiskey and beer. Other recurring Chinese purchases included potatoes, tea, various types of fish, baking powder, cabbage, onions, cloth, and matches.[10]

The store ledgers did not include Chinese-made products or Chinese-produced food and drink. The Chinese residents of Cortez clearly had other sources for things like soy sauce, liquor, dried or preserved fruit, medicine, tea, and opium. It was left to Chinese merchants to supply their community with goods imported from China via San Francisco. At least one was known to have conducted business in Shoshone Wells, and the census listed two other Chinese as storekeepers.

Thirty of the sites we investigated contained Chinese artifacts, but only twelve of the sites included five or more Chinese items. Of these, four had enough artifacts to be characterized as having a significant Chinese presence. This small number of sites did not reflect the overall Chinese population of the Cortez District, since the known Chinese population centers of Shoshone Wells, parts of the Cortez townsite, and the Garrison Mine were outside our archaeological investigation.

The facts of geography and population compelled the Chinese to live immersed in Western material culture, but the archaeological record shows the Chinese did continue some aspects of their traditional foodways. Stoneware storage jars and ceramic tableware were the most commonly found Chinese artifacts, despite the fact that Western equivalents like ceramic or metal plates, bowls, cups, and other tableware were easily available. Chinese stoneware jars held a variety of foodstuffs, including soy sauce and various preserves, as well as liquor, all manufactured exclusively in China.

This same Chinese tableware was also the strongest expression of Chinese culture in the archaeological record. The prolific use of symbols is a hallmark of Chinese culture, and one which the Chinese maintained in places like Cortez. For example, tableware decorated with flowers symbolized Confucianism, Taoist, and Buddhist principles. The Bamboo pattern symbolized virtue and longevity. The Eight Immortals design represented the patron saints of Taoism, transmutation, and happiness. The Four Seasons design stood for the pantheon of religion and philosophy; cherry equaled power; water lily equaled enlightenment and purity; peony equaled prosperity and nobility; and chrysanthemum equaled good luck. The Double Happiness design was used at weddings, symbolized happiness, and perhaps brought thoughts of spouses and families to married men on the other side of the world.[11]

FIGURE 6.2. Decorated Chinese porcelain: (a) Bamboo; (b) Celadon; (c) Double Happiness; (d) Four Seasons; (e) Attributes of the Eight Immortals. Courtesy of Summit Envirosolutions, Inc.

The Chinese and their lifeways did stand out at Cortez, but in terms of material culture non-Chinese artifacts always outnumbered Chinese artifacts, even at sites with a clear Chinese presence. Everyone found shelter in the same dugouts and rock or wooden houses, used the same tools for the same work, and shared the same space. They consumed much of the same food and drink, but often when the Chinese miner or mill worker retreated to his dugout for an evening meal he ate his own food—rice, fish, or dried fruit—from the same carefully decorated porcelain bowl he would use if he were still in China.

Native Americans

The Western Shoshone of the Cortez Mining District were both a distinct ethnic group and, uniquely, the original inhabitants of the region. The Euro-American explorers, fur trappers, and traders who first entered the area during the early 1800s set in motion profound changes in the Native American way of life. The tens of thousands of emigrants who followed, beginning in the 1840s, and the miners and settlers after them completely changed the Native American world.

Beginning in the 1700s, the Western Shoshone felt the effects of the Euro-American advances from all directions. American and British fur trappers and explorers came from the east and north, while Spanish and Mexican expeditions pressed in from the south. Horses, firearms, steel weapons and implements, and also deadly new diseases reached the Western Shoshone through trade and exchange long before they saw their first Euro-American.

In 1829, British fur trappers and traders led by Peter Skene Ogden worked their way west along the Humboldt River as far as the site of present-day Carlin, Nevada. The trappers hunted the river's beaver almost to extinction, but the later wave of Euro-American migration to Oregon and California proved even more devastating. The emigrants and wagon trains displaced the Western Shoshone from their land along the river. The newcomers' livestock devoured the seed-bearing native grasses that were so important to the Native food supply and generally disrupted the entire ecology. Some Native American groups eventually responded with violence, initiating a period of raiding and warfare that continued into the 1860s.

By 1863, as Andrew Veatch led his group of prospectors toward Mount Tenabo, the Western Shoshone had endured the loss of territory and access to critical seasonal resources, not to mention countless deaths from disease, hunger, and conflict. The Grass Valley and Crescent Valley corridor offered something of a refuge between the Humboldt River emigrant trail on the north and the Overland Route to the south. With the discovery of silver in the Cortez District, the Native Americans had no choice but to make uneasy adjustments even in this out-of-the-way place.

The Western Shoshone were at first observers and marginal participants in the Euro-American undertakings, beginning with the settlement in Mill Canyon. A newspaper report from May 1864 noted that a number of Western Shoshone were always there observing the work, although the writer's prejudice prevented him from appreciating their likely bemusement at watching the whites laboring on their mill or battering tunnels into solid rock.[12] But the Native Americans were losing their free access to natural resources and the water, grass-covered hillsides, root grounds, and stands of pinyon that fed them. They were frequently in desperate straits, and this often colored their relationships with Euro-Americans. One article noted, probably without exaggeration, that the Shoshone were willing to perform "any service required of them" in return for a crust of bread.[13]

Western Shoshone raiding, theft in the eyes of the Euro-Americans, shared the same motivation. This included breaking into isolated storehouses and occasionally confronting travelers on the road to Austin. At one time, Simeon Wenban had to dissuade a group of about sixty Native Americans from helping themselves to supplies at the Bullion Hill cookhouse.[14] An article in the *Reese River Reveille* described measures taken to protect stores of powder: "Under the roof of the shop is the door leading to an excavation in the hillside, where is stored a large quantity of powder. It is carefully guarded either from fire or the marauding propensities of the Indians, who much covet the article."[15] They frequently approached households asking for food or with offers of trade, exchanging pine nuts for beef, flour, or other commodities. The region's towns, like Austin and later Shoshone Wells and Cortez, also drew the attention of Native Americans. Not only were they sources of salvageable items, broken tools, discarded clothing, empty containers, and even food, but they also provided entrepreneurial opportunities for the sale of firewood, pine nuts, native grass cut for hay, and employment as domestic and farm laborers.

In the Cortez District, the Western Shoshone and the Wenban family enjoyed a good relationship. H. H. Bancroft stated in his biography of Wenban that the Native Americans knew him as "a kind friend, as well as an enemy not to be provoked."[16] According to Flora Wenban's diary, her father understood the importance of keeping local, personal conflicts from escalating into regional conflagrations. During the mid-1860s, when deadly skirmishes had broken out with the Tosawihi, or "White Knives" band of Western Shoshone, Wenban attempted to keep the Euro-Americans from turning indiscriminately against the Western Shoshone. He wrote to the Austin newspaper that there was no cause for alarm on the part of the sixteen residents of the Mill Canyon camp, as the Native Americans in their neighborhood had given every evidence of being friendly.[17] Years later, Wenban's granddaughter Flora Dean Hobart related the story of how one time about 1895 the family attended a midwinter fair at Golden Gate Park in San Francisco, which included an Indian village reenactment. Wenban recognized one of the players as a Western Shoshone he knew from Cortez, and he later had the man and his sister-in-law, also a member of the cast, as guests in their Van Ness Avenue mansion for several days.[18]

Native Americans were rarely employed in the mines and mills, although in May 1864 the *Reese River Reveille* reported that Western Shoshone laborers, paid with food, had built "Wilson's Grade" in Mill Canyon.[19] Simeon Wenban also hired Native Americans to collect salt from Grass Valley.[20]

The Western Shoshone established more reciprocal relationships with the area's ranchers and farmers, and they also took jobs at some of the businesses in Shoshone Wells and Cortez. The farmers and ranchers had monopolized much of the land and resources vital to the Native American way of life. Once they settled in, these same farmers and ranchers found themselves without competent, willing workers. They turned to the Native American men as hired hands and to Native women as domestics. They could not count on the miners, prospectors, mine and mill laborers, or other Euro-Americans to give up their occupations and become low-paid ranch hands. Even during mining downturns, these workers were more likely to simply move on to the next mining district.

Shoshone Wells was known to have a "camp" of Native Americans, which probably included women working as domestics in the Wenban household. In 1880, the US Census counted among the Western Shoshone living on ranches in Grass Valley five farm laborers and one "vaquero." There were also three women laundresses, one

FIGURE 6.3. A Western Shoshone woman
photographed at Cortez in the early 1900s.
She is seated alongside a traditional fiber
basket, a bucket made from an empty fuel can,
and a metal washtub. The long wooden objects
to her left are possibly pinyon poles. Courtesy
of Estelle Bertrand Shanks.

house laborer, one garden laborer, and three basketmakers.[21] Subsequent censuses through 1920 continued showing Native Americans as farm laborers, vaqueros, and domestic workers, including one washerwoman in a Cortez boardinghouse.

Domestic workers often enjoyed a special relationship with their employers and were intimate parts of the household. Flora Dean Hobart wrote that Mary Hall and her daughters, Western Shoshone women working at the Dean Ranch, "could take charge of a kitchen and cook for a crowd. They could equal a French laundress." She also said, looking back on the time, "Mary Hall is a matriarch and in fact I think my sister and I were in her matriarchy." Mary Hall would become a renowned basket-maker, at first making "the usual wash baskets" but going on to develop her craft and produce "more and more beautiful and finer" baskets. These were much in demand as works of art. She passed her skills on to her daughters, who were also celebrated for their basketry, beading, and buckskin work.[22]

There was, of course, another side to domestic work. One Western Shoshone woman described how "I used to cook for white ranchers. It was hard, trying work of long hours and low pay. At the end of the day I gathered up what scraps I could from the table and took them home so my children could have something to eat."[23]

Bill Englebright, a resident of the Wenban house during the 1920s and the son of Consolidated Cortez supervisor William Englebright, recalled that one of his earliest childhood memories had been watching an Indian woman digging on the hillside near their house. He asked his mother what she was doing, and his mother replied the woman was digging for roots.[24] The 1930 US Census included six Western Shoshone living in Cortez, among them the Maine family, who were descended from the first Chief Tu-Tu-Wa (or Toi Toi). They lived in a house near the school, which the children attended.[25]

Western Shoshone continued harvesting and collecting plant foods, particularly pine nuts, in the Mount Tenabo area throughout the twentieth century. Many of these locations were identified during interviews and field trips with Western Shoshone consultants during the early stages of our research.[26]

A special category of archaeological sites, referred to as ethnohistoric sites, represent the transition between the existing Native American culture and the new and dominant Euro-American culture. These sites are rare, and not always easy to define. They are characterized by either traditional Native American tools found among Euro-American artifacts, or traditional tools or other artifacts made from non-native material. For example, grinding stones, sometimes known to archaeologists by the Spanish *mano* and *metate*, for processing pinyon nuts or grass seeds, remained in use because there was no easily available Euro-American equivalent. Steel knives, on the other hand, quickly replaced their chipped stone counterparts. Glass has the same properties as obsidian or chert, and ethnohistoric sites often include artifacts such as arrow points or scrapers chipped from fragments of bottle glass. Trade goods are also potential markers of ethnohistoric sites, since the first interactions between Native Americans and Euro-Americans often consisted of exchanging furs or food for glass beads, buttons, trinkets, or tools like steel axes or knives.

Italians

Italians in the Cortez District were somewhat set apart, like the Chinese, but as time passed they became more established and integrated into mining district life. They maintained their own language and customs and often worked and lived together when engaged in particular trades or occupations, such as charcoal making. Censuses from the 1870s onward recorded numerous Italian families, the wives or new brides often immigrating well after their husbands had established themselves—a privilege not afforded the Chinese. The first Italians in the area came to Eureka in the 1870s, where they specialized in charcoal making. With the decline at Eureka late in that decade and into the early 1880s, many charcoal burners made their way to Cortez.

Italian names appeared frequently on Cortez company ledgers beginning as early as 1885. They included individuals listed as charcoal makers on previous census documents from the Eureka area. Newspaper stories also mentioned Italians, including one incident reported in the *Reese River Reveille* from May 1887. The article described a shooting in Cortez, in which Raphel Bianchi killed an Italian mill worker.[27] Neither man at the time was engaged in charcoal making. The victim, later identified as Giavanni Massera, worked in the Tenabo Mill. Bianchi made his living breaking and handling horses, driving a team, and working at other odd jobs, although at the time of the shooting he was a saloon keeper. Bianchi turned himself in to the sheriff in Austin and plead not guilty to murder. The ultimate outcome of the case is not known.

The 1900 US Census included six Italian households in the Cortez Precinct and two in the Garrison Precinct. Of the male individuals, eleven were day laborers and ten were miners. None were specifically enumerated as charcoal burners, although the "laborers" could have worked in the charcoal industry. In 1910, there were forty-three residents of Italian heritage in thirteen households. Again, none were employed in wood cutting or charcoal making, although by this time the conversion of the Tenabo Mill to cyanide leaching had all but ended the need for charcoal. In the 1920 census, John Rossi and his family were the only Italians in Cortez. Rossi, whose charcoal contract with J. C. Latta we saw earlier, was now a miner.

The frequent association between Italians and charcoal production has led archaeologists to sometimes conclude that charcoal ovens are signature Italian sites. Italians were not the only ethnic group involved in charcoal making. The Chinese, for example, manufactured charcoal in a number of mining districts throughout Nevada. The documented Italian involvement in the Cortez charcoal industry makes it safe to assume they were associated, perhaps not exclusively, in one way or another with our charcoal production sites. There are also "signature" Italian artifacts to confirm this connection. The most visible examples are representative of what might be considered an Italian diet. These include olive oil cans or jars, cornmeal or pasta containers, and snuff bottles. Snuff was a popular form of tobacco consumption among the Swiss-Italians, and one such artifact we found was an amber "wet snuff" bottle.[28]

Outdoor ovens constructed of rock or brick and used for baking bread were another indicator of Italian ethnicity. As archaeological features, they are found at historic logging camps, ranches, and railroad construction camps throughout the West. They are seen as a hallmark of Italian life, and in central Nevada they have been found associated with Italian ranches and farms as well as the charcoal workers of the Eureka District.

Breadmaking began with a fire inside the oven sufficient to heat the rocks of the structure white hot. The bakers then removed the coals, swept out the floor, sprinkled it with corn meal or flour, and put bread loaves inside to cook. The ovens could

FIGURE 6.4. The Berry children playing baseball in front of the Rossi house. The girl batting is Miola Rossi. Courtesy of Estelle Bertrand Shanks.

bake multiple loaves of bread, using up to fifty pounds of flour, within forty-five minutes to an hour.[29]

The Chinese, Native Americans, and Italians were the most recognizable ethnic groups in the archival records and archaeology of the Cortez District. But individuals from many groups had their own parts to play. Census records and other documents show Mexicans engaged in mule packing and mill work. An African American family ran a boardinghouse in Mill Canyon and included Maggie Johnson, who was part owner of the Silver Cloud and Treasure Hill Mines. Henry Berry, also African American, was a well-known assayer and worked for a time at Cortez.[30] Other nationalities found in the census, newspapers, and other documents included Canadians, Swedes, Finns, Germans, Swiss, English, and Portuguese.

LIFE IN THE CORTEZ DISTRICT

The people of Cortez wrested silver-bearing ore from deep under Mount Tenabo, extracted the precious metal, converted it to bullion, and sent it on its way to be minted into stacks of shiny new coins. From a distance, these people would be no more than tiny specks amid some humble buildings, walking or riding along a road or working at their trades. They contributed everything from brute labor to the knowledge of geology and chemistry needed to squeeze silver, an ounce at a time, from tons of rock. The written historic record tells us much about this cast of characters. Our archaeological study leads us further toward understanding what it was like to live in the Cortez District—to wake up there every morning, spend the day working, and then return home for the evening meal and a night's sleep. A story built from written words will always stay in the past, as we read those words in a memoir, store ledger, or census book, or even as we hear them in our minds as we read. But sometimes the gap between past and present disappears. You can hold a teacup from 1890 by its dainty handle, though it has lain buried for more than a century, just like the last person to use it did. You could even drink from it, providing it was not broken into a hundred pieces.

For years, the camp in Mill Canyon was the only significant settlement in the Cortez District. An 1863 Lander County census counted eighteen residents in Mill Canyon. These included a mix of the original locators—Wenban, Veatch, Cassell, McCurdy, McMasters—and other investors in the district's future mines and undertakings. In August, they organized a voting precinct in preparation for the September referendum in which Nevada voters began the process of upgrading their territorial status to statehood. The polling place was the office of the Cortez Mining Company, with D. Mullholland, Wenban, and T. D. McMasters as election judges.[31]

By May 1864, the Mill Canyon camp included three stone buildings and a log office building, all very substantial and neatly built, according to the *Reese River Reveille*. The two largest buildings were a dining hall and lodging house for the men employed in the mines. The company office was "large and convenient, [with a] fine library of scientific works chiefly relating to metallurgy, with bottles of acid, glasses, crucibles, and the carefully arranged cabinet of minerals." A stone blacksmith shop stood close to the mines, and a blasting powder storage room was excavated into the hillside, protected by a heavy door. A steam sawmill cut pinyon and juniper into usable lumber, although beams and timbers, or large quantities of dimensional lumber, still had to be freighted in from the Sierra Nevada pine forests. Other additions to the camp included a large, two-story frame hotel and a new saloon, the American, said to be "one of the neatest and best furnished saloons east of Virginia [City]."[32]

Simeon Wenban sent for his family in the spring of 1864. His fourteen-year-old daughter Flora documented the grueling, weeks-long journey from Ohio that included travel by riverboat, railroad, and finally stagecoach along the Overland Trail across Nevada. Wenban met them at the Ruby Valley station, more than 100 miles east of Austin. Flora said:

> When we were nearing Ruby Valley station the driver said to Mother, 'There is someone at the station who has been asking for you.' Of course she knew who that must be, and within a few miles of the station Father met the stage. After greeting mother he looked about for his children. Eva was sitting beside mother so he knew her, but I was sitting on the middle seat with another little girl about the same age and he didn't know which one was his, and was so excited he chose the other child as she was nearer Mother.[33]

The family continued on to Austin and from there traveled to Mill Canyon in an open wagon, lacking seats or covering. They spent a night at a ranch in Grass Valley and arrived in camp the next day. The Wenban family moved into the log cabin that served as the mining company office, complete with its mineral reference collection, and took their meals in the company boardinghouse. They were almost always the camp's only family, although Flora remembered once entertaining a mother and daughter also residing there. They largely created their own social life, passing time by playing cards and chess or making candy.

Other events typified life in a rough-and-tumble frontier mining camp. The *Reese River Reveille* of September 21, 1866, recorded a poker-inspired homicide in a saloon, perhaps the American, although the article described it only as being operated by a man named McGuire. One of the players lost not only his money but his pistol and Henry rifle. Claiming he was cheated, the loser shot and mortally wounded the winner and then immediately surrendered himself. The camp had no jail or law enforcement official, so the shooter spent the night in the custody of a prominent local citizen. The next day he was taken under guard to Austin, but he escaped along the way. He apparently seized his Henry rifle, which was being taken along as evidence, fired a few shots, and rode off.

The 1870 US Census showed forty-six residents in the Cortez District, residing in twenty-six households. Almost all were single men, and almost all of them worked at the mill or nearby mines. Fourteen men were mill hands, and eighteen, including Simeon Wenban, were miners. There was also a civil engineer, cook, and saloon keeper. Other professions included six Mexican mule packers and one teamster. In addition to Caroline Wenban and her daughters, two other women, described as "keeping house," lived in camp.

The population of Mill Canyon rose and fell in response to the need for workers which, in turn, depended on the success of the mines and mill. Hard winters and deep snow on the mountain's northern slopes and canyons sometimes shut down operations. At other times work stopped, and people left the camp because their employers ran out of money, ore stockpiles were insufficient—or excessive—or the mill shut down for repairs. Flora Wenban described how, when the camp emptied out, "the place was very quiet." For recreation, they climbed the nearby mountains and admired the "long views." For two kids coming from the farmlands of Ohio, the mountains must have been an exciting diversion.[34]

At the same time, a smaller settlement appeared near the head of Cortez Canyon, west of Mount Tenabo at some springs the Western Shoshone had dug out and incorporated into their seasonal travel. It became known as Shoshone Wells, although the Cortez District's original claim book referred to it as "Mugginsville." It had six residences in the spring of 1864.[35]

Shoshone Wells grew in importance during the 1870s, as mining shifted to the St. Louis, Garrison, and other locations near the Nevada Giant. A colony of Chinese miners took up residence in a series of dugouts near the Garrison Mine, and they would have made regular visits to Shoshone Wells to buy supplies, socialize, or send mail or other communications to the outside world. Shoshone Wells was a mile-and-a-half walk from the Garrison, which probably discouraged workers from making it their home. In 1886 the new town of Cortez began drawing away Shoshone Wells' population, however, it still included at least 50 residences in the years between 1872 and 1910.[36]

A century later, in the early 1980s, the University of Nevada began a multiyear archaeological research project that delved into life at Shoshone Wells.[37] The Shoshone Wells site covered an area about the size of a large city block. The archaeological remains included dugouts, stone and adobe buildings, and wood frame structures clustered in small neighborhoods, some mixed and some with ethnically distinct populations of Chinese, Italians, Euro-Americans, and Mexicans.

Dugouts were the most common dwelling in Shoshone Wells. These began as U-shaped excavations into the natural slope, with stacked rock forming the interior walls. The front and side walls were made from upright logs, milled wood, or rocks. Roofs were flush with the top of the back wall, and either shed or low gable style. They were covered by a layer of earth over juniper poles, canvas, recycled metal, or flattened cans. There was often a stone hearth in a rear corner, with a chimney or metal vent pipe extending through the roof.

The freestanding stone structures were mostly small, single-room buildings made with unmodified cobbles gathered from the immediate area. Walls were caulked with mud and small stones. Shoshone Wells also included wood frame buildings, but their rock foundations were all that remained at the time of the study. A number of adobe buildings had survived a century of abandonment, although the eroded mud blocks showed only traces of the plaster and whitewash that once covered the interior walls.

Along the camp's main street, the University of Nevada archaeologists identified numerous adobe buildings, dugouts, and foundations. Artifacts from theses residences showed both Euro-Americans and Chinese living in the neighborhood. Chinese inhabited a separate group of seven adobe buildings, called the "bottomland cluster," from as early as 1872 to about 1910. This neighborhood also included a number of foundations arranged around a deep dugout that may have served as a small temple. It would have been the place where the immigrant Chinese could practice and maintain their religious and social customs. An additional group of dugouts, also inhabited by Chinese, extended along a stream channel at the north end of town.

Italian woodcutters lived in a group of four stone houses on the hillside above town. The lack of Chinese artifacts in this area, coupled with abundant Italian wine bottles and more common food cans, clued the archaeologists into the ethnicity of the tenants. The neighborhood also included an outdoor oven—a cherished, traditional cooking feature from Italy.

FIGURE 6.5. The Wenban home, mid- or late 1880s. The view is south toward Grass Valley, with Shoshone Wells in the background. Courtesy of Bureau of Land Management, Battle Mountain District Office.

The University of Nevada investigation gave us a picture of life for the large group of Chinese miners who comprised a thriving little community in Shoshone Wells during the 1870s and 1880s. They resided in dugouts and adobe and wood frame houses in two adjacent neighborhoods, one clustered around the temple and along the main road through camp, and a second along the draw to the north. According to the Lander County tax rolls, at least two Chinese-owned stores served the needs of the Chinese population.

The person most responsible for the prominent position of the Chinese in the Cortez District, Simeon Wenban, also made a home for himself and his family at Shoshone Wells. We know from Flora Wenban's diary that the family lived in Mill Canyon when they first arrived in 1864. The 1870 census counted them among the residents of the "Cortez District," which, judging from the miners, mill workers, mule packers, and teamster listed along with them, almost certainly meant the settlement in Mill Canyon. The Wenbans moved to Shoshone Wells sometime during the late 1870s, which put them closer to their mines and the future site of the Tenabo Mill. The family hosted the wedding of their daughter Eva to local rancher Joe Dean on February 12, 1878, at their Shoshone Wells home.[38]

The Wenbans' residential and administrative complex was well established by 1885, when Lander County tax rolls listed the property as a "dwelling house, office, store, and barn." People described the Wenban home as a small mansion, with all the Victorian accents proper to a man of his stature. It had twelve-foot ceilings, redwood siding, and a landscaped lawn enclosed by a white picket fence. There was a parlor

FIGURE 6.6. The Wenbans' parlor. The pictures, light fixtures, furniture, carpet, and fabric crowding the room represented the common Victorian practice of displaying wealth to visitors, who were directed to this room upon entering. The comforts the Wenbans enjoyed in this remote mining camp were unlike anything anyone else in the district experienced. Courtesy of William Magee.

for greeting and entertaining guests, and the glorious luxury of running water, piped in from a nearby spring.[39]

The Wenbans' Victorian home was no palace, especially compared with their later Van Ness Avenue mansion in San Francisco. But it was easily the most impressive residence in Shoshone Wells, befitting the wealthiest, most prominent family in the district. It was torn down during the 1950s, reportedly to salvage the wood for building material. The University of Nevada archaeologists found little in terms of artifacts or refuse in the immediate area. The foundation was a low stone wall, forming a rectangular enclosure 56 feet long and 23 feet wide, with a stone-lined cellar at one end. Broken wood and fragments were the only traces of the home's elegant plaster cornices, redwood siding, red brick chimney, and shake roof.

The archaeologists discovered the Wenbans' household dump about 250 feet away. It included plenty of evidence of Victorian conspicuous consumption, appropriate to the family's elevated social status. The quantity and proportion of "luxury" artifacts were much higher than at family households elsewhere in Shoshone Wells. These included decorated ceramics, such as imported French porcelain from manufacturers like Limoges, Haviland, and Sarregueminies. There were also cut crystal, etched glass, and personal items like a glass collar stud and carved mother of pearl. Children's toys included a tricycle and cat's eye marble.[40]

The Wenbans, like the families of other Nevada mining magnates of the day, kept strong social, business, and financial ties to San Francisco, including real estate investments. In 1869, Flora and Eva were sent to there to school, although they returned home for the summer. Their first trip in 1869 was arduous, as the Transcontinental Railroad was still under construction across Nevada. They traveled to Austin on George Russell's stage, spent the night, and headed west on the Overland Stage. Another stage took them north to intersect the completed portion of the railroad, apparently at Winnemucca, where they finally caught the train to San Francisco. When they returned home in June of that year, with the railroad finished, they took a passenger train to Beowawe, at the time nothing more than a telegraph office and a railroad worker bunkhouse. Simeon Wenban met his two daughters with a picnic, which they enjoyed together on the banks of the Humboldt River.[41]

Flora Wenban married William Ogden Mills, the nephew of Darius O. Mills, whose Bank of California was a major Comstock financial institution. The couple remained in the area, and Mills died in Cortez in 1891.[42] Their two children, Carrie A. Mills (born 1887) and Simeon Wenban Mills (born 1889), were both born in Cortez.[43] Eva and Joe Dean had two daughters, Flora and Ethel. Eva became a widow in 1885 when her husband was killed in a dispute with a local sheepherder.[44] She married William P. Shaw in 1890.[45]

Wenban purchased a lot on Van Ness Avenue in San Francisco in 1887 for $40,000, and erected a "magnificent home."[46] His biographer said he "removed" to San Francisco in 1888 to give his family, including four grandchildren, "all the advantages of metropolitan life."[47] The extended family continued living in the Cortez District, but eventually they divided their time between Cortez and California. Wenban kept a close eye on his mining affairs, as well as Nevada politics. He and his wife Caroline traveled to Washington, DC, in 1889 as part of a Nevada delegation to President Harrison's inauguration.[48] He stayed personally involved in Cortez as principal owner and manager of his mining operations. In his absence, they were watched over by "faithful agents."[49]

In 1895, six years before his death, the *San Francisco Call* carried a thumbnail description of Simeon Wenban. The writer noted Wenban was presently not working as much as when his mines "were being sought for by everybody": "Mr. Simeon Wenban, the capitalist, passed into the Palace Hotel yesterday, stepped lightly across the asphaltum cortel and disappeared into the billiard room, puffing his cigar and looking for all the world like a contented millionaire."[50]

THE ARCHAEOLOGICAL VIEW

Throughout the latter part of the nineteenth century, people in the Cortez District lived in Shoshone Wells or made homes in the new town of Cortez, near their work in the Nevada Giant mines and the Tenabo Mill. Others spent their time in smaller camps, or alone, on the slopes and foothills of Mount Tenabo. The products of their labor—cord wood, charcoal, lime—linked them to an undertaking reaching from deep within Mount Tenabo all the way to the Transcontinental Railroad at Beowawe. They worked and lived within sight of the mines and mill that had made Simeon Wenban a rich man. Some hoped for a silver strike all their own; others mined or prospected for wages. They practiced trades to service the district or worked to keep it supplied with fuel and raw material. But whether they were part of a woodcutting crew, hired miners, or independent prospectors, their day-to-day existence could not have been more different than the life of a "contented millionaire."

As we explored the everyday lives and work of people living in the Cortez District, two sites figured prominently in our investigation. These sites—which we called The Prospectors' Camp and A Chinese Enclave—contained mining features, stone and earth buildings, outdoor ovens, and numerous artifacts. They dated from the 1880s to the turn of the twentieth century, a prosperous time for the district that began with construction of the Tenabo Mill and closed with Wenban's death. Neither of the sites, nor their tenants, appear in historic documents or photographs. Everything we know about them comes from our archaeological study.

The Prospectors' Camp

The site was a small mining enterprise, located partway up the steep, rocky slope of Mount Tenabo, with a view out over Crescent Valley. To a handful of miners, it was a promising location worth exploring, and they invested enough time and labor to excavate four prospect pits and three adits.

They lived in two stone structures built of rock collected from the surrounding area. They stacked the rock without using mortar, to form walls about a foot and a half thick. One building was about twelve by twelve feet, with posts at the corners. The other was comprised of two adjoining rooms and measured about nine by twelve feet.

When we began work within the single-room structure, we first encountered the remains of the roof. After it was abandoned, the building gradually deteriorated until the roof, beams, and other supports collapsed into the interior. Years of wind and slopewash then covered everything with a layer of soil. We found enough intact roof to determine that it was at least partially made with milled wood, probably scavenged from somewhere else in the district. The roof lay almost directly on the floor, with a small gap between them that helped protect and preserve some of the artifacts.

The dark, hard-packed floor of a historic structure represents many years' accumulation of dirt, food scraps and spills, broken bottles, and empty cans, all intermixed with the small, personal items that tend to get lost, or swept into a corner, or pressed under a miner's boot. It is the debris of life, deposited without anyone

FIGURE 6.7. Chinese stoneware jar, padlock, and miner's candle holder. Courtesy of Summit Envirosolutions, Inc.

noticing, and containing everything from lost buttons and coins to whatever the last inhabitants happened to leave behind the day they abandoned the building.

About half a dozen men lived in the two structures, judging from the size of the rooms. They most likely prepared meals and socialized in the single room, in front of the fireplace, since this is where we found the most and widest variety of artifacts. These included food cans and medicine bottles, as well as odd individual items, like a padlock, and even a single human tooth. The tooth was a curious find, and we can only speculate on how it ended up in the floor. Perhaps it was the result of a fight or freak accident, or simply needed to be pulled. The other building contained less evidence of day-to-day activities and may have been a dormitory or sleeping rooms.

The mix of artifacts from the single-room structure showed the miners enjoyed no separation between their work and social lives. They brought the tools of their trade home, perhaps to safeguard them but also for repairs, sharpening, and other maintenance. The artifacts we found comprised a miner's personal tool kit. Included was a handmade candleholder made out of a short piece of iron rod, one end split and bent into hooks for fixing it in cracks or crevices in the rock. The other end of the rod was forged into a loop to hold the candle. Other mining artifacts included drill steel, a long bolt converted to a rock drill by shaping the threaded end to a point, and a broken ore bucket handle. There was also a brown glazed Chinese stoneware jar, which would have been a useful container but did not by itself mean the miners were Chinese.

Additional work-related artifacts included several black powder cans, a smaller can for percussion caps, and a variety of hardware items. Wire mesh screen, carriage bolts, nails, chain link, and an axe head were testament to the tasks undertaken

FIGURE 6.8. Medicine bottles, including a Dr. Hostetter's bottle, and a hard rubber irrigator. Courtesy of Summit Envirosolutions, Inc.

by the workers that lived here. Horseshoes and nails indicated the miners had draft animals on hand for transportation or heavy hauling.

Life's domestic side was reflected in the several lost buttons and clasps, all from various articles of clothing that would have been laundered on the metal washboard found on the site. Coffee pots, a plain white ceramic mug, numerous bottles, and tin cans told us what the miners ate and drank and that they had access to numerous consumer products from well beyond the boundaries of the mining district. Other artifacts told of life's discomforts. The miners used Dr. Hostetter's Stomach Bitters as a supposed cure-all for internal pains, although its 47 percent alcohol content probably had a lot to do with its curative reputation. Other medicines were administered with a black, hard rubber irrigator that we found on the site.

Canned fruit, vegetables, meat, and fish were the norm in the district, with fresh fruit and vegetables a rarity. The miners cooked with lard, and numerous baking powder cans pointed to biscuits as a common element of any meal. They drank beer, wine, and ale, and fragments of animal bones showed they ate modest amounts of chicken, mutton, and beef. Imported French pickles added some variety to the diet. Isolation, and maybe a lack of cash, meant recycling and repurposing whatever was at hand. Tin cans, for example, were made into strainers by punching holes in the bottom. Handles were attached to cans to make cups, and others became small buckets with the addition of a wire bail.

A Chinese Enclave

This site was a workers' camp on the west side of Mount Tenabo, about half a mile south of the Prospector's Camp. While that site was clearly mining-related, the Chinese Enclave did not include mining features or enough tools or other artifacts allowing us to draw conclusions about the residents' main occupation. Among

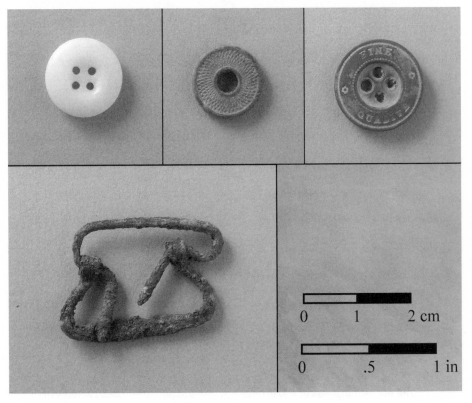

FIGURE 6.9. Buttons and a suspender clasp. Courtesy of Summit Envirosolutions, Inc.

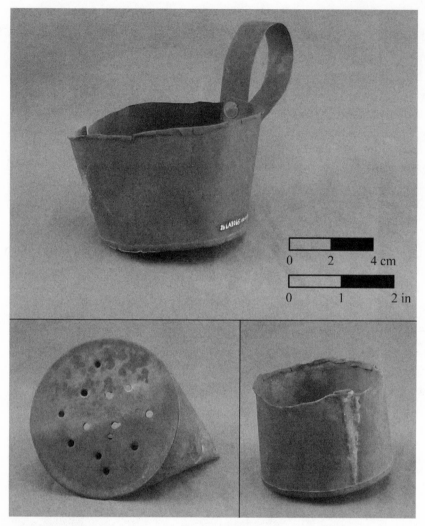

FIGURE 6.10. Tin cans modified for use as strainers and cups. Courtesy of Summit Envirosolutions, Inc.

thousands of artifacts, we found two axes, one broken handle from an ore bucket, and some horseshoe nails. The occupants could have been miners, woodcutters or charcoal makers, or perhaps hostlers, feeding or caring for pack mules or freight teams—or all of the above. You could have found workers engaged in any of these occupations within a short walk from the site. One thing was clear, however—at least some of the inhabitants were Chinese.

The site included six rock structures and one tent platform. The structures were partially dug into the slope, with low rock walls extending out to form the enclosures.

The walls were mostly dry-laid, although some had traces of natural dirt mortar. Wood poles and posts formed a superstructure supporting the roof. The buildings were small, about ten to twelve feet on a side, with the exception of one dugout that measured fifteen by twenty feet. Each had its own fireplace, which indicated they functioned as residences rather than storage or work structures. They were large enough to accommodate a number of individuals, and we estimated from sixteen to twenty people lived in the camp. One of the structure's fireplaces was equipped with a steel lintel and had attached hooks for suspending cooking pots. This might have served as the camp's dining and social center.

The six structures were built sometime during the 1880s and 1890s. One pinyon support post gave a tree ring date of 1881. The floor of another structure yielded a number of coins, among them an 1883 nickel, an 1895 quarter, a 1902 half dollar, a 1911 dime, and several Chinese coins.

Archaeologist love coins, though not for their monetary or collective value. Coins have dates, and those dates give us at least one exact time when the site was occupied. The coins were all found together, and because they stay in circulation for many years this meant they were deposited, or lost, sometime after 1911. The 1911 dime provided the year after which the artifacts from that particular portion of the floor must have been deposited. Archaeologists refer to this date as the *terminus post quem*. Other evidence suggests the building was constructed long before, perhaps the early 1880s. We do not know if it was continuously occupied or whether various groups who came and went with the fortunes of the district simply took advantage of existing shelter.

The coins' value added up to 90 cents. While we might think of that as pocket change, in 1911 it was the equivalent of about $18 in today's money. The person who lost these coins was probably very displeased. The Chinese coins were all several hundred years old and had long since stopped functioning as currency. They were likely brought from China as gaming pieces or talismans or used as buttons.

Four of the dugouts included enough varieties of Chinese artifacts for us to conclude the occupants were Chinese. These included items from day-to-day life, such as rice bowls, teacups, distinctive Chinese cans, medicines, and opium paraphernalia. We also found an intact lock of black hair, looking very much like a *queue*—a long braid of hair worn by Chinese men. Various Euro-American items were intermixed with the Chinese artifacts, showing the Chinese had also adopted, by choice or necessity, what was available.

The diet included canned fruit, vegetables, fish and meat, evaporated milk, lard, baking powder, and dry ingredients for biscuits and other baked or prepared foods. We found beef, sheep, and pork bones and bones from several fish species. Meat cuts included leg of mutton and lamb; rib racks and cured pork shoulder; and beef sirloin, rump cut, rib, and chuck roasts. These were probably broken down and made into stew or roasted whole on a spit, and then eaten by the group at communal meals.

FIGURE 6.11. An unassuming depression and wooden posts (*above*) concealed a stone dugout and fireplace buried under three feet of sediment. Summit Envirosolutions, Inc. Photo by Robert McQueen.

FIGURE 6.12. Coins from the Chinese Enclave: a 1911 dime (*top*); an 1895 quarter (*middle*); and a 1902 half-dollar (*bottom*). Summit Envirosolutions, Inc. Tangerine Design & Web.

FIGURE 6.13. Euro-American items from the Chinese Enclave included a tin bowl, enameled cup, and coffee grinder. Courtesy of Summit Envirosolutions, Inc.

Cost analysis of the beef cuts indicated the residents did not settle for cheap meat but could afford mid- to higher-priced cuts.

Opium has long been associated with the Chinese sojourners of the American West. It was common in nineteenth-century China, so it comes as no surprise that Chinese workers would bring use of the drug with them to America. The work was hard and the men were a long way from home. Opium relieved aches and pains and perhaps sent the smoker into a dreamlike state far from a lonely mountainside camp in a strange country. Archaeologists consider opium paraphernalia—pipes, tins, and lamps—as indicators of Chinese presence at a site. The Chinese Enclave included abundant and varied opium paraphernalia. One structure included opium containers and smoking apparatus, leading us to think it might have served as the camp's smoking "den."

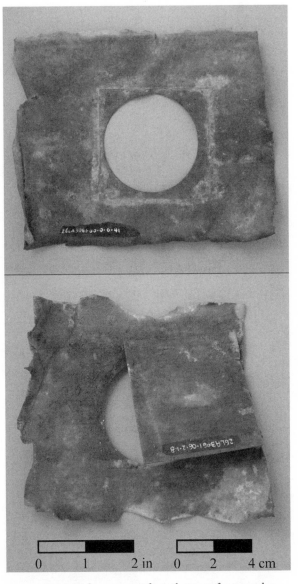

FIGURE 6.14. Metal squares cut from the tops of rectangular cans. The square patch welded over the round hole mark the cans as Chinese, probably containing tea or other dried foods. Courtesy of Summit Envirosolutions, Inc.

FIGURE 6.15. Euro-American whiskey and patent medicine bottles. Wright and Taylor brand whiskey was found at several sites in the Cortez District. While apparently popular with residents of all backgrounds, its prevalence suggests that the district's one company store supplied everyone's needs. The small bottle on the lower right contained Chinese medicine. Courtesy of Summit Envirosolutions, Inc.

月　残　吹　更　三

高　胡　水
告　琴　東

FIGURE 6.16. A terra cotta opium pipe bowl. It would attach to a hollow wood or bamboo pipe shaft. The characters translate to *late at night, all is dark, only the moon provides very little light.* (Translation courtesy Priscilla Wegars, Curator, Asian American Comparative Collection, University of Idaho.) Courtesy of Summit Envirosolutions, Inc.

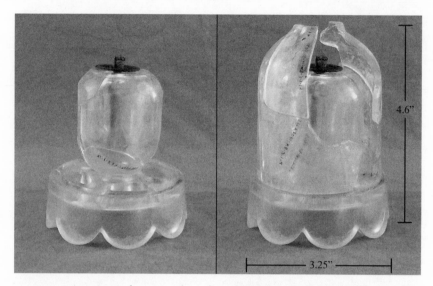

4.6"

3.25"

FIGURE 6.17. A rare, complete opium lamp from the Chinese Enclave. It is comprised of four parts: the stand and oil reservoir (*left*); the chimney (covering the reservoir at *right*); and the wick holder attached to the top of the reservoir. The wick holder is a Chinese coin. The lamps are thought to be English-made, shipped to China, and re-exported to the United States or anywhere with an immigrant Chinese population. Courtesy of Summit Envirosolutions, Inc.

The Town of Cortez

Put yourself on Mount Tenabo in 1881, looking down from your charcoal camp, or dugout, or over the backs of your mules. You would see the Garrison Mine, or at least the growing pile of waste rock spilling out from its portal and down the mountain. You might see pack trains making their way up the trail and disappearing into Arctic Canyon. Wisps of smoke would mark the charcoal kilns—easy to spot in islands of bare ground among the trees. A few dugouts would sit empty while their inhabitants labored invisibly in mines and prospects.

Then come back in five years. A sprawling, outsized mill would fill the area at the foot of the slope below the mines, distinguished by a stately brick smokestack. The other addition to your view would be the mismatched scatter of homes, boarding-houses, hotels, and businesses that had made space for themselves in the sagebrush on the alluvial slopes below the mill.

The town of Cortez developed in tandem with the Tenabo Mill. Wenban took this new town into account when he designed the water system to supply the mill and serve the needs of a small community. Naturally, miners and mill workers wanted to live close to their workplaces, and with them came their families and others involved in the many supporting occupations necessary to keep the district going.

FIGURE 6.18. Sixteen opium cans were found on the
site. They were made of brass, which helped preserve
the original paper labels. The last few letters of opium
can be seen in the circular remnant of the customs
stamp in the top view. The Chinese characters
identify the company name and quality of the
product. Courtesy of Summit Envirosolutions, Inc.

FIGURE 6.19. Northern half of the Cortez townsite, with the Tenabo Mill in the fore-ground. The white, two-story building at the center of the photograph is the boarding-house, set among the scattered residences and other structures comprising the town. Courtesy of Northeastern Nevada Museum.

Census data and other estimates[51] had the population of the Cortez District hovering around 250 to 300 people during its most productive years from the mid-1880s through the 1900s. Although we lack census figures for 1880 and 1890 it is fair to say the majority of inhabitants lived in the Cortez townsite. (The 1880 census might have shown the relative populations of Shoshone Wells, the Garrison Mine area, and Mill Canyon, but the Cortez District schedules have been lost. The entire 1890 US Census was ruined by fire and subsequent water damage at a government warehouse in the 1920s and was then discarded.) The 1900 census counted 274 residents, divided between the Garrison Mine Precinct in Eureka County and the Cortez Precinct in the Lander County portion of the district. The county line split the Cortez townsite, but we assume the inhabitants of the town comprised majorities in each precinct. We can also extrapolate backward from 1900 because, even with ten years' difference, the same mine, mill, and surrounding community were in place by 1890.

A close examination of the 1900 US Census shows a much more diverse population than the earlier Mill Canyon camp, which we will discuss further in later chapters. The most obvious contrast between the town of Cortez and the male-dominated mining and prospecting outpost in Mill Canyon was the addition of families. By 1900, fifty-six women and forty-nine children resided in the combined Cortez and

Garrison Precincts. Of approximately sixty-five total households, thirty-two could be described as family units. These most commonly included a husband, wife, children, or other relatives but also couples without children. Occupations were overwhelmingly mining or milling related, such as "miner of ore," "mine laborer," or "mill laborer." Supporting occupations indicative of a more complex community included merchants and store clerks, hotel keepers, blacksmiths, butchers, physicians, barbers, laundresses, and seamstresses. The 1910 census reflected much the same family and gender composition, but the closing of the Tenabo Mill and the change to individual lessees in the mines reduced the town's population to about sixty people.[52]

Cortez grew quickly, but it remained just as much an outsized mining camp as a real town. At one time or another it had a school, post office, water system, and, during the 1920s, limited electricity, but no church, newspaper, or municipal court, sewerage disposal, or trash collection. The townsite was a stark, lonely place chosen because it happened to be the closest relatively level, open area to the mines and the Tenabo Mill. The "town" was a spontaneous collection of buildings and dirt roads. Most residents lived in dugouts or in a series of wood frame and adobe houses aligned along one or two roads. There were fenced yards here and there, including locust trees and lilacs that survive today, but for most residents the sagebrush began one step outside the door.

Many families rented houses owned by the Tenabo Mill and Mining Company, while single men stayed in the large, white company boardinghouse or the hotel across the street. Some people had gardens, and some raised their own poultry and livestock, but goods were limited to whatever was available at the company store. Life included few extravagances. Folks spent most of their money on basic bulk commodities, with an occasional splurge on something like "Epicurean lobster," a Thanksgiving turkey, or fashionable fabrics from the East.[53]

Our investigation did not include the central Cortez townsite, but we explored a number of residences and numerous dumps on the outskirts of town, or strung out along roads leading away from town.

Town Dumps

A number of dumps were located on the southern edge of Cortez, on sloping land cut by a shallow ravine. A short distance away, a scattering of similar dumps pointed toward some structures in the south part of town. The inhabitants of these buildings, or their neighbors, were the likely source of the refuse. Together, the dumps spanned more than thirty years' time. One, dating to around 1904, represented the early twentieth century. Two dated from 1915 to 1930, and one was from the 1930s. These piles of discarded items provided glimpses of life from three distinct times during Cortez history. The first marked the years immediately after Simeon Wenban's death, when the Tenabo Mill shut down and lessees took over mining. The second

FIGURE 6.20. A typical roadside dump on the outskirts of town, including a number of multiple-serving food cans. Courtesy of Summit Envirosolutions, Inc.

originated during the Consolidated Cortez operations, and the third represented the Great Depression.

Refuse in the dumps likely came from a number of different households in town, but the artifacts can be compared to show the similarities or differences in the way people lived and how this might have changed over time. In all three periods, food cans made up the majority of artifacts, along with bottles and bottle glass fragments. Cortez residents relied heavily on canned foods—fruit, vegetables, meat—and this did not change from the end of the nineteenth century through the Depression. The high-altitude climate, a short growing season sandwiched between late spring and early autumn frosts, and limited water ruled out commercial truck farming. Fresh fruit and vegetables had to come by rail to Beowawe, and then either wagons or auto-trucks took them the rest of the way to Cortez. Evaporated milk took the place of fresh milk, as shown by the hundreds of condensed and evaporated milk cans found in the dumps. It did not need refrigeration and was a basic ingredient for biscuits and other casserole-type dishes. Meat was always part of the diet, with beef and chicken bones found in the early dump, and beef, lamb, and pork bones as well as numerous eggshells from the Depression-era dump. Other domestic and personal items also remained constant, with all the deposits including items like teacups, tableware, metal pans, and buttons and clasps.

The two earliest deposits marked a commonly observed change in tin can manufacturing, with the 1904 dump including a mix of the older style hole-in-cap food cans and the newer "sanitary" cans. Sanitary cans, basically the modern crimped seal tin can, came into common use just after the turn of the century and replaced the older style can by the 1920s. They were considered "sanitary," or safe, because they were sealed without the use of lead solder, an acknowledged health hazard even then. At Cortez, the transition was more or less complete by about 1915, as only sanitary cans were found in the 1915 to 1930 dumps.

The 1904 dump contrasted with the others in that it seemed to represent a simple residential household, while artifacts from the later dumps reflected a more complex living situation. Personal items from the 1904 site included such things as porcelain doll parts, a rivet snap from a satchel purse, pieces of children's shoes, and a porcelain teapot spout. The later refuse deposits had similar background household artifacts, but they included an additional mix of mechanical and plumbing artifacts not seen in the earlier dump. These included water pipe, pipe couplings, iron rod, automobile and wagon parts, carriage bolts, and an oil can, along with various pipe fittings, bolts, and nuts. There was also chicken wire and slag or clinkers.

The contrasting artifacts could have been from different households or from residents with different occupations. But they also held clues about how life at Cortez changed between 1904 and the Consolidated Cortez period. The camp in 1920 was a shadow of the 1890s and early 1900s Cortez, and Consolidated Cortez had to upgrade and expand the water system to serve the workers the company would need once the mines and new mill went into operation. It is possible the new residents had to manage hooking up to the water system themselves and, jumping ahead to the 1930s, the remaining residents were on their own with respect to drinking water after the mill closed. As one resident during this time recalled, "every once in a while when something happened they'd get in their little pickup and away they'd go across there [Grass Valley] to fix it, get it patched up so we had good water."[54] Artifacts from the later dumps showed at least one household did their own plumbing, while other residents regularly worked on automobiles and wagons. They also burned coal, which would have come from the Beowawe railhead, as well as raising their own chickens.

Valley Dwellings and More Town Dumps

One of the main roads into and out of Cortez was marked by the remains of a number of structures—two above ground and one double-room dugout—as well as a series of small dumps that were clearly examples of institutional cleanup and disposal.

Table 6.1. Selection of cans found in one of the several town dumps. An example of how archaeologists organize the information we collect. The data can then be compared with other features and sites and with sites from other research projects. Notice, for example, the large number of evaporated milk cans and the predominance of one type of tobacco can.

TYPE/SHAPE	PRODUCT	STAMPED/MAKER'S MARK	MODIFIED?	DATE RANGE	COUNT
Hole-in-cap	Meat	"LIBBY'S/VEAL LOAF/PORK & MEAT BY-PRODUCTS/U.S. INSPECTED & PASSED/ESTAB. 22"	No	1906–1930s	1
Hole-in-cap	Meat		No	1899–1930s	1
Hole-in-cap	Condensed milk	"BORDEN'S CONDENSED MILK" on one	No	1903–1908	2
Vent hole	Evaporated milk		No	1917–1929	66
Vent hole	Evaporated milk	"PUNCH HERE"	No	1935–1945	18
Vent hole	Evaporated milk		No	1915–1930	36
Vent hole	Evaporated milk		No	1945–1950s	141
Three-piece can	Fish	"FRANCE"	No	1866–ca. 1888	2
Stamped end	Fuel/kerosene		Yes	1865–1920s	3
Stamped end	Paint		No	1847–1985	1
Sanitary	Spice		No	1904+	1
Sanitary	Stew	"STEW/1205"	No	1904+	1
Sanitary	Beer	"BREWED AND PACKED.../H[AMM'S]/SM[OOTH]" and "...BREWING ST. PAUL, MINN."	No	1935–1950	1
Sanitary	Beer	"BEER CAN . . . / BOTTLE . . . / . . . BREW . . . / . . . CONTENT . . . / . . . INTERNAL . . ."	No	1935–1950	1
Sanitary	Coffee	"RADIANT ROAST"	No	1910–1960s	6
Sanitary	Tobacco	"SATISFACTION GUARANTEED/WILL NOT BITE THE TONGUE"	No	1905–1988	57
Sanitary	Potash	"B. T. BABBITT'S/PATENT DATE APRIL 4th 1882/PURE POTASH"	No	1904+	2
Sanitary	Indet.	"REMOVE/PROTECTIVE CAP/PUNCH BOTH HOLES"	No	1904+	1
Drawn	Fish	"NORVEGE/D"	No	1880–1933	2
Drawn	Fish	"NORVEGE/D" on one, "KIPPERED HERRINGS/PACKED IN NORWAY/NORVEGE" on one	No	1897–1933	7
Drawn	Fish	"MONTEREY CAL. USA [in circle]; "2H14B/31FM"	No	1918–unk.	1
Drawn	Fish	"KIPPERED HERRINGS/PACKED IN NORWAY/NORVEGE"	No	1933–1950s	6
Drawn	Tobacco	"OLD ENGLISH/CURVE CUT/PIPE/TOBACCO"	No	1870s–unk.	6
Slip lid	Percussion caps	"CALIFORNIA CAP CO/SAN FCO"	No	1880–1905	1
Slip lid	Tobacco	"COPENHAGEN/SATISFIES"	No	1911–unk.	10

A Wood Frame House

The standing remains of a fireplace and chimney marked what we knew was some kind of structure, probably part of the new and expanding 1886 Cortez townsite. Once we began, it became clear we were excavating the remains of a wood frame building. We exposed the decomposing remains of a wood floor in the structure, clearly an improvement from the dirt-floored dugouts that sheltered so many Cortez residents. At least one resident was female, as indicated by the coin purse, nail brush, thimble, and hair pin we found within the structure. The occupants were modestly well off, as analysis of beef bones showed they were from comparatively expensive cuts of meat. A single, expensive style of Chinese rice bowl reinforced this social standing, although without indicating whether any residents were actually Chinese, well-off or otherwise.

Food-related evidence pointed to a varied diet, which added to the sense of a better, more interesting lifestyle than we found at some of the workers' dugouts higher up on the mountainside. This included cooking staples—baking powder, lard, and evaporated milk—but we also found seeds and shells from seasonal fruits and nuts, and a variety of meats including lamb, beef, pork, and chicken. Beverages included whiskey, beer, wine, soda water, and fresh milk.

A Saloon and a Cortez Dugout

We suspected this otherwise ordinary building was a saloon when we uncovered its stone foundation and found an inordinate number of shattered liquor bottles. The structure was built shortly after the founding of Cortez in the late 1880s and was occupied until about 1910. The bottles appeared to have been thrown out a doorway, where they piled up against the wall. Whether the liquor was sold here or not, drinking it was clearly a major pastime at this spot.

The dugout consisted of two rooms cut into a small hill and sharing a common middle wall. It was built after 1890 and inhabited up to about 1900. The room comprising the east half of the dugout yielded buttons, clothing rivets, suspender clasps, footwear, and clothing fragments. The west room had glass from more than a dozen bottles, can fragments, eggshells and butchered bone, an 1890 dime, a metal pan or wash basin, and a fireplace. The distinct contrast among the artifacts indicated some degree of organization to the distribution and use of space. The east room served as sleeping and residential quarters while the west room was for cooking, eating, drinking, and socializing. We found the same separation of the "dormitory" from meal preparation, consumption, and socializing at the prospectors' camp farther up the mountain.

All Kinds of Dumps

A number of other dumps and refuse deposits were found along the road between Cortez and Shoshone Wells. We could not tie them unquestionably to specific

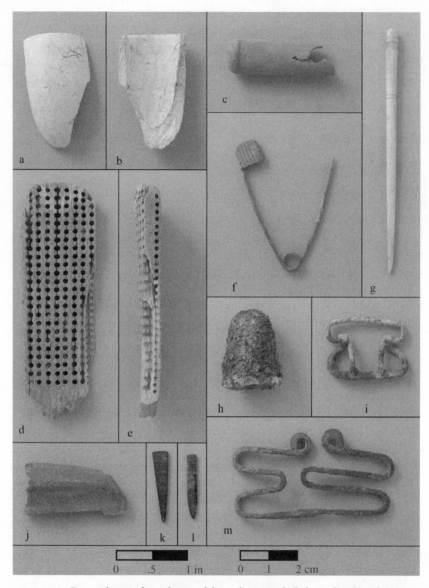

FIGURE 6.21. Personal items from the wood frame house included pipe bowls and stems (*a/b*, *c*), nail brush (*d*, *e*), safety and hair pins (*f*, *g*), thimble (*h*), comb parts (*j*, *k*, *l*), and a clothing fastener (*m*). Courtesy of Summit Envirosolutions, Inc.

households, but they still added significant details to our knowledge of life at Cortez. The various cans and other containers held products from all over the United States, as well as France, Norway, Scotland, and England. Common name brands included A. Booth (East Coast oysters), Armour Packing Co. and N. K. Fairbanks Co. (lard,

FIGURE 6.22. A mug and plate from the dumps. These are examples of plain, durable ceramics known as hotelware, used in boardinghouses and other commercial establishments. Courtesy of Summit Envirosolutions, Inc.

canned meat), Royal Baking Powder, Schillings Best (spices, extracts), Folgers (coffee, tea), and Log Cabin syrup. There were mining artifacts like assay crucibles, spikes from ore cart tracks, and an ore cart wheel. A pair of sheep shears told us the occupation of at least one Cortez tradesman. The dozens of buttons we found were mostly from work clothes. Chinese artifacts included five opium tins and a number of storage jars and other ceramic fragments.

The things people throw away tell us a lot about their lives, and overall refuse patterns are also revealing. We determined that some of the secondary dumps along this mile-long section of road between Cortez and Shoshone Wells came from a commercial-scale dining hall, like the Cortez Hotel or the boardinghouse. These were distinguished by large, approximately one-gallon, multiserving food cans and numerous evaporated milk cans. They frequently included "Dutch Cleanser" brand scouring powder cans, clinkers and unburned bits of coal, and tobacco tins. There were also condiment bottles, ink bottles, aspirin tins, lamp oil cans, a broken electric light socket, parts from a kerosene lamp, an egg beater, a clock spring, accordion keys, a toothpaste tube, cuticle cream, and a bottle of dandruff cure.

The multiserving cans came from an institutional setting where large amounts of food were prepared at one time. Evaporated milk would have been used in baking, probably biscuits or simple breads to go with rich, hearty meals. The Dutch Cleanser cans found at each dump indicated frequent and repeated cleanup, also on a commercial scale, since no single family would use up that much cleanser.

The dumps represented repetitive cleanup and disposal from the same place, a task that produced piles of similar debris every time. Each one comprised about a wagon or small truck load. Someone, probably a kitchen helper, occasionally took wagons of trash a few hundred yards from town, pulled off into the sagebrush, and threw it out. The refuse's relatively tight time frame spanned the two decades between 1904 and the revival of large-scale mining and milling by Consolidated Cortez in 1923.

During this time, the Wenban estate leased its mines and claims to numerous small operators, who worked sporadically and opportunistically, focusing on the quickest and most easily available ore. Such lessees were even less likely than the earlier prospectors, miners, and mill workers to become permanent residents. The large, white boardinghouse so prominent in the photographs of Cortez would have offered the flexible living arrangements required by uncertain, small-time mining. It was quite possibly the source of the loads of trash dumped along the road to Shoshone Wells—an otherwise innocuous act that became a boon to our archaeological study.

NOTES

1. *Reese River Reveille*, July 25, 1863.

2. Wegars (1993); Dirlik (2001).

3. Bancroft (1889, 18).

4. *Reno Evening Gazette*, June 10, 1881.

5. *Reno Evening Gazette*, November 10, 1896.

6. Angel (1881, 429).

7. In 1890, a portion of the 1890 Eleventh Census of the United States was badly damaged by fire, although the original population schedules were unharmed. These were stored in the Commerce Building in Washington, DC, but in January 1921 another fire and subsequent water damage rendered the 1890 Census virtually unusable. No efforts were made to salvage the records, although they were kept in storage until 1935. Then, for reasons which are still unknown, they were destroyed. See Blake (1996).

8. 1900 United States Federal Census, Garrison Mine Precinct, Eureka County, Nevada; Cortez Precinct, Lander County, Nevada.

9. McElrath (1998).

10. Johnson and McQueen (2016, chapter 105: Consumer Behavior and Material Culture).

11. Brott (1982).

12. *Reese River Reveille*, May 7, 1864.

13. *Reese River Reveille*, July 25, 1863, 1.

14. Magee (2010).

15. *Reese River Reveille*, May 5, 1864.

16. Bancroft (1889, 249).

17. Magee (2010, 23–43); *Reese River Reveille*, May 31, 1865.

18. Hobart (1954).

19. *Reese River Reveille*, May 5, 1864, 1.

20. Murbarger (1959, 12).

21. 1880 United States Federal Census, Grass Valley, Lander County, Nevada.

22. Hobart (1954).

23. Patterson, (n.d., 8).

24. Englebright (2011).

25. Shanks (2012–2015).

26. Rucks (2000).

27. *Reese River Reveille*, May 9, 1887.

28. Praetzellis and Praetzellis (2009).

29. Johnson and McQueen (2016, chapter 104: Demography, 283–284).

30. Harmon (2011).

31. *Reese River Reveille*, August 15, 1863.

32. *Reese River Reveille*, May 5, 1864.

33. Magee (2010, 33).

34. Magee (2010).

35. Hardesty (1988, 81, citing Cassel 1863 [Cortez Mining District Claim Book], 25, 257, and *Reese River Reveille*, April 7, 1864).

36. Hardesty (1988, 84).

37. Hardesty and Hattori (1982; 1983; 1984).

38. Woolley (1999, 165).

39. Hardesty (1988, 87).

40. Hardesty (1988, 87).

41. Magee (2010, 43).

42. *Daily Nevada State Journal*, September 11, 1891.

43. Bancroft (1889, 259); 1900 United States Federal Census, San Francisco, California, Enumeration District 227, Sheet 12.

44. Bancroft (1889, 259).

45. Woolley (1999, 167–177).

46. *Nevada State Journal*, October 22, 1887; Bancroft (1889, 259).

47. Bancroft (1889, 259).

48. *Daily Territorial Enterprise*, March 12, 1889, 3.

49. Bancroft (1889, 259).

50. *San Francisco Call*, October 11, 1895.

51. Bancroft (1889, 255).

52. Johnson and McQueen (2016, chapter 104: Demography of the Cortez District).

53. Johnson and McQueen (2016, chapter 105: Consumer Behavior and Material Culture).

54. Shanks (2012–2015).

Chapter 7

Beyond Food, Clothing, and Shelter

Our archaeological excavations at Cortez uncovered evidence of what people ate and drank, what they wore, and what kind of shelters they lived in. The next step, building on this evidence, was to expand our view to take in the population as a whole and the society in which people lived. Ethnic identity aside, who were they? What kind of place did they make of this faraway mining camp, or, more to the point, what kind of place were they *trying* to make? How successful were the residents of Cortez in imposing the ideals and aspirations of Victorian ideology, which held sway at the time over American society, on this harsh and demanding spot on the mining frontier?

AGE, GENDER, AND FAMILY

As we have seen, 250–300 people lived in the Cortez District during its best years. Archaeological evidence suggests a higher number. The dugouts, tent flats, and stone buildings on the slopes and foothills of Mount Tenabo and on the outskirts of Cortez could have accommodated anywhere from 300 to 400 people.[1] It is quite possible the census enumerators passed over many of these isolated camps. Either way, the population expanded dramatically from the 18 people who lived in Mill Canyon in 1863, or the 46 counted during the 1870 census, to several hundred inhabitants by the mid-1880s. In 1889, H. H. Bancroft estimated at least 300 people lived in the district,[2] and this roughly correlates with census data, although the official counts fall short of Bancroft's minimum. The 1900 US Census counted 274 residents in the Cortez and Garrison Mine precincts. The population dropped to about 100 in 1910, 40 in 1920, and remained between 30 and 40 through the 1930s and into the 1940s. At least a couple of hundred people lived in Cortez during the Consolidated Cortez period in the 1920s, although this bump in population did not show up on the decennial censuses. Consolidated Cortez had not yet reopened the Garrison and Arctic Mines in 1920, nor had the mill been built. In 1930, they had just closed those same mines and mill, depopulating the district weeks before that year's census.

Young men, either single or unaccompanied by their wives and families, made up the lion's share of the initial population of the Cortez District. They either shared living quarters or stayed in boardinghouses. As the district evolved, some left empty-handed while others established families or sent for their absent wives and children. Men remained the core of the workforce throughout the life of the district, laboring as miners and mill hands. Some residents owned and worked their own mines or claims. Other men, and a few women, filled the other necessary occupations, from painters and carpenters to cooks and merchants.

As the Cortez District stabilized and developed, the number of families went from two in the first Mill Canyon years to thirty-two in 1900. There were three women in the Cortez District in 1870, and fifty-six in 1900 in the combined Cortez and Garrison Precincts. Almost all were housewives, however a small number had occupations such as laundress, dressmaker, schoolteacher, and midwife. One woman, Maggie Johnson, was a mine owner.

The 1900 census counted forty-seven children, eighteen of whom were attending school. At least thirty children were born in Cortez between 1887 and 1916, and reports for the 1907–1908 school year showed twenty-four students in the Cortez school (making it fourth largest in Lander County).[3]

Recognizing the presence of women or children at an archaeological site is straightforward, in theory. While most artifacts enjoyed common usage by all ages and genders, certain things were more likely owned or used by women or children. Obvious examples would include jewelry, women's clothing, makeup and cosmetic containers, purses, garter clasps and clips, corset stays, and sewing implements. The last two illustrate the uncertainty involved in linking specific artifacts to specific genders, since men also wore corsets as back braces, and even bachelors had to sew and repair clothes. Artifacts that hint at a woman's presence, but are less definitive, include finer ceramics and eating utensils and nicer home furnishings—the things an itinerant bachelor typically did without. Children's artifacts included toys (balls, dolls, wagons, tricycles, dishware, marbles), clothing (buttons, shoe fragments, pins, clasps), and baby buggies.

Our study identified twenty sites with artifacts commonly associated with women or children, and eleven of these had artifacts related to both. The sites dated from the 1880s to the early 1900s, when the district was most successful and the population at its highest. Research in other mining districts found that women and children most often lived either in the main camp or town or within easy reach. Our study area focused on the Mount Tenabo slopes and foothills, away from the population centers at Cortez and Shoshone Wells, but we did confirm this distribution in the sense that the many work camps we investigated in the outlying areas were almost exclusively male domains. When we did identify women or children in the archaeological record, those sites were usually close to or along a major road. This gave women and children relatively easy access to town, within walking distance of the schools, stores, and entertainments important to family life.

A VICTORIAN MOMENT

American society of the late nineteenth and early twentieth centuries was bound up in Victorian ideology, which set social standards and dictated individual behavior,

FIGURE 7.1. Cortez miners, ca. 1889. Many of the men are holding candles, and several very young children and young boys are included in the photograph. Courtesy of Northeastern Nevada Museum.

even in places like the Cortez Mining District. Victorian ideology was marked by a compulsive work ethic; punctuality; a worldview dominated by rationality, order, and natural laws; morality; temperance; the cult of domesticity (genteel behavior and upward mobility); knowledge and self-cultivation; cleanliness; and conspicuous consumption of mass-produced, brand-name goods. Victorian ideals not only stressed public education but advocated continued education beyond the early grades.[4]

The Victorian class structure with its inflexible social categories and primacy of birthright and family connections never had an easy time in America. The basic premise of a gold rush literally trampled the idea of a rigid, hierarchical social order underfoot. An unknown, penniless prospector could be moments away from the discovery that would make him a millionaire and, like Simeon Wenban, open the doors of high society and political influence to himself and his family. Nevertheless, people knew their place. Wealth and privilege were not equally distributed, and it was the Victorian sensibility that often governed the expression of class differences in day-to-day life. Material possessions were one such expression—left behind for archaeologists to discover and interpret.

Victorianism at Cortez was broadly visible archaeologically in the layout of settlements, the pattern of refuse disposal, and the content of the well-removed dumps. Nothing reflected Victorian conspicuous consumption more than brand-name goods, including food and drink and the bottles and cans they came in. These were, according to the values of the time, collected and disposed of outside the settlement. Of course, this ideal was unevenly applied, and we found no shortage of refuse closely associated with individual residences and households.

CLASS IN THE CORTEZ DISTRICT

A typical mining camp class structure consisted of a lower class of semiskilled and skilled workers; a middle class of white-collar professionals and merchants; and an upper class of mining capitalists, engineers, and managers. At Cortez, lower class occupations included day laborers, woodcutters, and more skilled workers such as

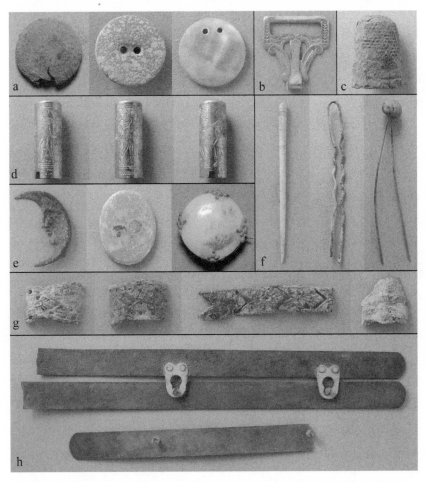

FIGURE 7.2. Women's artifacts recovered at Cortez: (*a*) buttons; (*b*) garter clasp; (*c*) thimble; (*d*) lipstick tubes; (*e*) pin/brooch; (*f*) hair pins; (*g*) bracelet; and (*h*) corset stay. Courtesy of Summit Envirosolutions, Inc.

miners and millers, blacksmiths, carpenters, and charcoal burners. Middle class workers included professionals and business owners, specifically saloon and board-inghouse keepers, clerks and bookkeepers, physicians, teamsters, assayers, and schoolteachers. Engineers, mill and mine foremen, and of course mine owners comprised the upper class. The largest group by far in the Cortez District was the working class. Our archival research showed that throughout the district's history lower class workers made up 80 percent or more of the population. Middle class professionals and merchants varied from only 4 to 17 percent, understandable because a small mining camp like Cortez offered very limited commercial opportunities. The upper class never included more than a handful of households.

A tenet of archaeological study holds that, not surprisingly, wealthier, higher class individuals and families would own more expensive, higher-quality goods. We found it useful to focus on three class- and cost-sensitive artifact groups: ceramics (tableware, cups, serving dishes, etc.); meat cuts, as shown by discarded bones; and personal items. We were able to classify four sites in our study as relatively high status, with the great majority of others falling into the working class category. The distribution of these sites followed the pattern recognized at other mining districts, with the higher-status sites located closer to town while lower-status groups lived in outlying camps.

Ceramics

Victorian households marked their status through the use and display of increasingly fancy or more expensive ceramics, with the highest-quality decorated types often being imported. As we pointed out, University of Nevada archaeologists recovered expensive, imported ceramics in refuse from the Wenban house. In addition, the disposal of refuse well away from the house was an example of Victorian sanitation practices. In our research, we compared the relative percentages of decorated ceramics to percentages of plain ceramics and other, cheaper kinds of tableware, such as tin plates. One typical upper class site about half a mile northwest of the Cortez townsite included—among numerous decorated ceramics—porcelain, ironstone dishes, and a tortoise-shell-shaped, intricately decorated sugar bowl or condiment bowl lid. There were also a corset stay, garter clasp, and other women's artifacts, indicating the site was not only high-class but a family residence as well.

Meat Cuts

Along with china and silverware, the meat a person dines upon says a lot about their social status. Beef cuts are especially useful to archaeologists because, first of all, bones often preserve well and, second, beef in one form or another was part of almost everyone's diet. Different cuts of meat varied considerably in desirability and

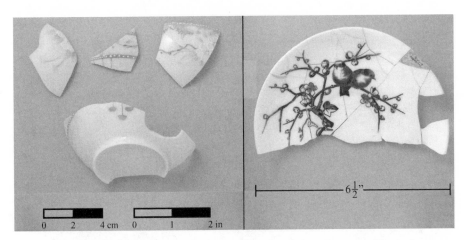

FIGURE 7.3. Expensive, highly decorated ceramic bowl and plate. Courtesy of Summit Envirosolutions, Inc.

cost. Higher-status families and individuals could afford more desirable meat, which we recognized from the bones we found. Less expensive cuts included roasts, ribs, and leg portions, which, after butchering and consumption left discarded vertebra, lower leg bones, shoulder blades, and pelvic bones. The majority of sites included low- to moderate-ranked cuts of meat, which supported our general conclusion regarding the working class nature of the Cortez population. The four higher-status sites included bones from pricier cuts.

Personal Items

We can also determine status by examining the amount and quality of personal possessions, such as clothing and footwear, jewelry and adornments, grooming and cosmetic products, medicines, and what we would call indulgences: alcohol, tobacco, and opium. Individuals can, of course, own a few special, overly expensive personal things, but they can combine with other indicators to distinguish high status sites.

Alcohol and Tobacco

The second half of the nineteenth century saw a general, national decline in alcohol consumption. One reason was the strong temperance movement supported by Protestant women, which was in turn an offshoot of the Victorian ideals of moderation and sobriety. But alcohol and tobacco use was as endemic to the mining frontier as its eschewing was to Victorian sensibility. Victorians had no choice but to concede on the issue, so long as it was limited to males drinking in saloons. A proper Victorian woman simply did not consume alcohol in public.

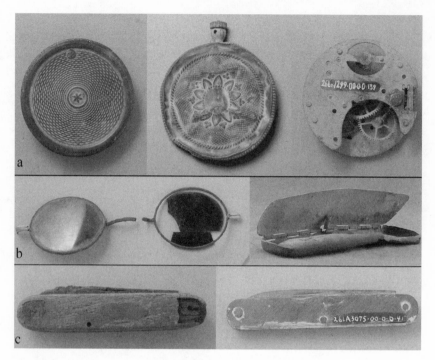

FIGURE 7.4. Personal items from our excavations included pocket watches, dark glasses and case, and pocket knives. Courtesy of Summit Envirosolutions, Inc.

While Victorian convention publically scorned drinking and the use of tobacco, it was another story behind closed doors. We found alcoholic beverage containers, totaling almost 400 bottles, at 41 sites. Alcohol was consumed at nearly every residence in the district, with beer being most common, followed by equal proportions of wine or champagne, whiskey, and other liquor. Five sites were probably drinking houses or saloons, as they contained more than half the beverage bottle total.

Sites with only male inhabitants generally showed less evidence of alcohol consumption. The large number of alcoholic beverage bottles from the handful of saloons indicated men did their drinking in a social setting, in town. Houses where families lived, that is with evidence of both males and females, tended to have more evidence of at-home alcohol consumption. This may have reflected the Victorian prohibition of public drinking by women or that married men, many of them fathers, did not participate in their bachelor counterparts' public drinking.

Twenty-nine sites contained tobacco-related artifacts such as pocket tobacco tins, tobacco tags, fragments of smoking pipes, cigar cans, one glass humidor lid, and a brass lighter. These were found in nearly all the large dumps, but few came from individual homes, probably because empty cans and pocket tins were discarded along with the household refuse.

Health and Hygiene

To the Victorian mind, good health and hygiene were inseparable. We discovered a range of grooming and hygiene-related artifacts in our excavations. These included tooth brushes and tooth powder, hair tonic, combs and brushes, razors and other shaving paraphernalia, skin lotions and creams, lipstick, hand mirrors, and shoe polish. Household items like cleansers, scrubbers, brooms, mouse traps, and animal poison exemplified domestic hygiene. Late nineteenth- and early twentieth-century Cortez residents either made use of home remedies, or they bought proprietary medicine from the company store. These "medicines" were sold without a prescription and, contrary to popular belief, were not patented. That would have required revealing the contents, something the producers wanted kept secret. Instead, manufacturers copyrighted product names, designs, and even the shapes of bottles to prevent infringement or copying and still avoid government oversight. We also recovered a number of medical devices, including fragments of at least six glass syringes and a hard rubber irrigator identical to an ear syringe depicted in an 1895 Montgomery Ward catalog.

Consumption

One distinction between the working class and Cortez's few upper class individuals was the latter's ability to travel more frequently and to special order goods and materials for delivery to the camp. But living in this isolated mining district constrained even the well-to-do. Everyone went to the company store and made their selections from the same limited supplies, which may have inadvertently acted as

FIGURE 7.5. Bottles and jars containing the various medicines, cosmetics, and cold creams available to people at Cortez. Courtesy of Summit Envirosolutions, Inc.

a social equalizer. We examined several ledgers from the Tenabo Mill and Mining Company store, from three different periods: 1886 to 1890, 1900, and 1912 to 1913. These records were an important aspect of our work because they described perishable items that, because they decay, are mostly invisible to archaeological study.

The purchases covered the basics: noodles, barley, rice, cornmeal, flour, sugar, potatoes, and baking powder. Beverages included coffee, tea, and whiskey, while cheese, milk, butter, and eggs made up the dairy products. Much of the milk came from dairies around Reno, and the eggs arrived from Utah. There were apples, preserved jellies, and meats, including beef, ham, pork, and bacon. Much of the beef came from local ranches, while the pork came via the railroad, salted and packed in barrels. Chicken was notably absent from the ledgers, probably because people raised them locally for their own consumption and for sale or trade to their neighbors. Canned meat included chipped beef, deviled ham, and corned beef. Fresh produce came from California's Central Valley, especially the Sacramento area, and included seasonal tomatoes, peaches, plums, pears, string beans, cantaloupe, watermelon, cucumbers, grapes, cabbage, and strawberries.

The most commonly purchased clothing included overalls, shirts, boots, shoes, socks, gloves, jumpers, and suspenders, all typical worker's attire. Needles, thread, buttons, and cloth were less common. It appears that the male clientele could buy ready-to-wear items in Cortez, but women either made their own clothes or, on rare occasions, special ordered dresses and other apparel. There was little variation in the available merchandise over the twenty or thirty years from the 1880s and the last ledger entries in 1913. Overalls bought for $1.25 in 1888 cost the same in 1913.

Even though the residents of Cortez and its surrounding areas were from diverse backgrounds, the company store's limited stock kept them from expressing ethnic differences through their choices as consumers. The Chinese were an exception, as they bought clothing, food, and other specialty items from Chinese merchants. These restrictions also dampened manifestations of class or upward mobility. Conspicuous consumption may have been in the back of everyone's mind, but the ledgers and artifacts we recovered tell us functionality and durability were the watchwords, and only rarely were purchases made that could prompt admiration or envy. Work pants were the long-wearing Levi Strauss jeans; buckles and other fasteners were plain; footwear consisted of galoshes and work boots; and even most pocketknives lacked embellishment.

NOTES

1. Johnson and McQueen (2016, chapter 104: Demography of the Cortez District, 243).
2. Bancroft (1889, 255).
3. Ring (1909).
4. Schlereth (1991); Hardesty (2010); Brown (1968).

Chapter 8

Archaeology and Memory

The most personal form of history is memory. We used everything from old news-papers to bits of broken glass to assemble a story of life in the Cortez Mining District. We dug deep into musty ledgers and deep into the hard-packed dirt floors of lonely dugouts. But as our story moved from the distant past toward the present day, a new resource—memory—came into play. We not only had the good fortune to meet and interview two people, Estelle Bertrand Shanks and Bill Rossi Englebright, who spent parts of their childhood at Cortez, but we were also the beneficiaries of interviews, transcripts, and accounts made by others with the wisdom and foresight to recog-nize and preserve history in its most fragile and fleeting form.

We start with the US Census of 1910, where we meet Charlotte "Lottie" Allen, age sixteen. She lived with her family in Cortez and, as the census sheet notes, worked as a "restaurant waitress."[1] Many decades later, Charlotte's son Frank and her daugh-ter Estelle, along with her brother Clarence Allen, recorded their memories of life in Cortez as part of the University of Nevada's Lander County Oral History Project.[2] Estelle also contributed many additional hours of conversation to our study. Another name from 1910 is a seven-year-old Cortez resident named Miola Rossi. Eleven years later she would marry William Englebright, the newly designated superintendent at Consolidated Cortez. Their son Bill spent the first few years of his life with them in Simeon Wenban's house in Shoshone Wells, and he personally contributed his family history to our research.[3]

The census described Charlotte Allen as a waitress, but according to her family the job meant being an all-around maid, cook, and housekeeper. It was work she not only mastered but returned to many times during the course of her life. Her par-ents, Charles and Estella Allen, moved to Cortez from Eureka, when the mines there closed in 1907 or 1908. Charles Allen worked for the Tenabo Mill and Mining Com-pany as a blacksmith and foreman. His son Clarence told how his father's workday consisted of sharpening picks and drills first thing in the morning, then going to the mine to supervise the crews. He worked year-round, never taking a holiday, even for Christmas, and earned $4.50 a day.

Charlotte's daughter Estelle remembered her mother describing the "education" she got in running a hotel. She did everything there was to do at one time or another, for $15 a month. It "went pretty far," as she said, since her employers provided free room and board. Still, she had to save for a couple of months to buy a new pair of shoes or books of poetry. Whittier, Longfellow, and James Whitcome Riley were her favorites.

FIGURE 8.1. Charlotte Lottie Allen, about age sixteen. Courtesy of Estelle Bertrand Shanks.

Charlotte Allen gave a newspaper interview in 1977 in which she described her work in the small mining camps that sprang up all over Lander and Eureka Counties in the first decades of the twentieth century.[4] The boardinghouse kitchen crew got up every morning at 4:00 AM and began making the day's twenty-five loaves of bread and two pans of rolls from dough that had been prepared the night before and left to rise. The miners arrived for breakfast and ate bacon, sausage, and hotcakes by the dozen. Their lunches, also made the night before, included at least two thick sandwiches and pie or cake. Once the breakfast tables were cleared, tubloads of sheets and linens waited to be washed and ironed. More cooking followed, on a scale that was by no means easy. For example, Charlotte mashed potatoes by hand in a kettle three feet high. Later on she had her own boardinghouse, where she provided room and board for eight to twelve miners, who never hesitated to complain about what she was charging them.

In 1912, at age nineteen, Charlotte married a miner named John Hamlyn, who was boarding at the hotel where she worked. He was, as Estelle recalled her mother's description, a cultured man who enjoyed listening to classical music on the phonograph records of the time. He died six years later during the 1918 Spanish flu epidemic at Hilltop, a small mining camp in the Shoshone Range between Cortez and Battle Mountain.

Charlotte, now widowed, returned from Hilltop to Grass Valley, near the family ranch. The Allens moved to the ranch when Charles's job at the mine ended, about the time Consolidated Cortez took over the Tenabo Mill and Mining Company. The 1920 census showed Charlotte sharing a residence with her fifteen-year-old cousin, who had himself been orphaned by the flu epidemic. She soon went back to work at the boarding house in Cortez. In November 1920, she married Matthew Bertrand, a miner who took his meals at the boardinghouse.

Bill Rossi Englebright, the son of William Englebright and Miola Rossi, also shared with us stories and memories of his family's time at Cortez. John Rossi, his wife Lena, daughter Miola, and her four siblings lived in Cortez in 1910. Rossi was born in Italy about 1870 and immigrated to the United States where his family settled in Eureka in 1874. He was a miner in 1910, but as we have seen in Chapter Five, he also had a history in the charcoal business.

In 1920, with the Cortez population down to thirty-five, John Rossi worked as an independent silver miner. The three Rossi children still at home included Miola, now seventeen. William Englebright probably came to Cortez to assess the Wenban properties in connection with their purchase in 1919 by Consolidated Cortez. He was born into a mining family in 1889 in Nevada City, California, and graduated as a mining engineer from the University of California in 1913. He would have had a mutual mining interest with John Rossi, although at the time Rossi was hauling ore to the railhead at Beowawe, not working underground. William Englebright inevitably crossed paths with the Rossi family in the small community and met his future wife.

William Englebright and Miola Rossi were married on December 23, 1921. According to family history, William felt obligated to pay off one of his deceased father's bad mining investments, but the couple married once the debt was discharged. Because of his position as general mine superintendent with Consolidated Cortez, the Englebright family took up residence in the Wenban house.

The year 1924 was important for the Bertrand and Englebright families. Estelle Bertrand was born in February, and Bill Rossi Englebright was born in November. The Bertrands lived in Beowawe at the time, and Charlotte traveled to Reno to give birth. Two weeks before their baby was due, the Englebrights temporarily relocated to Elko. Married to a mining superintendent or not, Miola worked in the hospital for those two weeks until her son was born.

William Englebright and his family spent the better part of the 1920s in Cortez, living in easily the most comfortable home in the district. But Simeon Wenban's Victorian sanctuary did not set them apart from mining camp life. William chopped his own firewood and, as the story goes, a slip of the axe once resulted in a serious wound to his ankle. With the nearest hospital or doctor miles away, he sutured the cut himself. On another occasion, Miola fell ill with strep throat one winter. When her temperature approached 106 degrees, her husband decided he had to act. He drove her by car to Beowawe, where they stopped the first passing freight train and rode in the caboose to the hospital in Elko. Doctors removed Miola's tonsils and resolved the infection.

The Englebrights left Cortez for Canada about 1928. Charles Kaeding, an associate from California and general manager of the Cortez project at its start in 1923, convinced William to move to Toronto and join him in some Canadian mining ventures. William Englebright went on to enjoy a long career in mining development, with projects in locations as far flung as British Columbia and the Orinoco River in South America. He died in Toronto in 1948.[5]

The Bertrands were more typical of the families who made a living moving from job to job in the small camps, mines, and mills of north-central Nevada. The 1910 census gave Matthew Bertrand's occupation as "saloon keeper"; he was a miner when he met Charlotte Allen; and he worked at Cortez as a pump tender during the 1920s. He also managed the post office and Consolidated Cortez company store in Cortez, before spending the 1930s at various mines and mills in the region. At the Tenabo camp, in the Shoshone Range about twelve miles northwest of Cortez, he operated a tractor hauling ore and supply wagons to and from Beowawe. As his brother-in-law Clarence "Doc" Allen described it, he ran "a big old tractor that hauled ore, pulled wagons loaded with ore, from Tenabo to Beowawe."[6] This was probably similar to the "traction engine" the *Reese River Reveille* described in 1907,[7] which brought merchandise, supplies, and equipment from Beowawe and returned pulling trailers loaded with ore.

The Bertrands moved from Beowawe to Cortez in 1927, when Matthew was hired as caretaker and pump minder at the Grass Valley well and pump station. We do not know what mark the Englebrights left on the archaeological record at the Wenban house, but we do have a more readable connection between the Bertrands and the pump station. Our work at the site had revealed much about the pre-Consolidated Cortez, steam powered operation of the pumps and water system. But we were also able to separately explore its archaeology from the 1920s and beyond, including the Bertrands' time there.

The first personal description of the pump station is from the steam-powered era, in a memoir by Mabell Paddock McElrath (born in 1892).[8] Her father was superintendent at the Tenabo Mill during her childhood, and she remembered two visits with him to the pump station. One time the pump man took her to his small garden and let her select lettuce, green onions, and radishes to take home to her mother. Even as a child, she noticed the pump man had water for his garden although it was a precious commodity normally reserved for the mill. On another occasion, she accompanied her father on an inspection of the pump house, and the Chinese pump man invited her for lunch. They ate their rice and vegetables and drank tea together on the dirt floor, indoors, beside a large water tank.

The Bertrand family lived at the pump station from 1927 to 1929. Both Estelle and her older brother Frank have left us their child's perspectives of life there.[9] For three-year-old Estelle, their arrival was a moment of sleepy confusion. The family had driven from Beowawe and did not reach their destination until well after dark. Estelle woke up in the car, in the darkness, and began to cry. She told her mother she wanted to go home. Her mother answered, "Well, this is our home now."

FIGURE 8.2. The Bertrand children and their father at the pump house residence. Courtesy of Estelle Bertrand Shanks.

There were two buildings at the pump station, one housing the pumps and the other for the caretaker and his family. Frank Bertrand remembered they lived in a small, "regular house" that was fairly comfortable, especially considering the dugouts and drafty frame structures available to most mining camp residents. The company power plant provided the added benefit of electricity. The single bedroom included their parents' big bed and two small beds for the children. They bathed in a big No. 10 washtub, heated on the wood-burning cookstove and then moved onto the floor. Blankets hung on chairs screened the tub and provided a measure of privacy. They rinsed their hair from a kettle of warm water.

Frank's chores included chopping wood and collecting eggs in the chicken house, which Estelle remembered as a sod-roofed dugout—a mysterious, dark, and quiet place for the children to go. Their mother raised rabbits and chickens, and they often had visitors who walked down from Cortez to buy them.

As for the pump house itself, Frank said: "I was around the big equipment quite a bit, the pumps and what not with my father. I never did get in the wells, they scared me. It was quite an experience, but there was just my sister and I, Estelle, and we played together and roamed around through the sagebrush."[10]

The pump station was vital to the successful operation of the mill and to life in the town. The pump minder had an important responsibility and needed to be competent, experienced, and independent enough to keep things working no matter what problems cropped up. It was not uncommon, according to Estelle, for her father to spend nights sleeping on a cot in the pump house when the pumps were in full operation.

We had already uncovered what amounted to a small industrial complex at the pump house site, complete with concrete machinery mounts, and the remains of structures going back to the 1880s. The pumping area included a mix of cut and wire nails spanning the period between 1886 and the 1920s, as well as insulators, fuses, and light bulb glass marking Consolidated Cortez's electrification of the operation. There were a few fragmentary Chinese artifacts, possibly evidence of a single individual, like the pump minder who shared his lunch on the dirt floor of his dwelling with Mabell McElrath in the early 1900s.

But what of the "little house" the Bertrands lived in? There were no standing structures at the location when we began work, but we found its remains on a rectangular flat or platform about 50 feet northwest of the pump complex. We assumed the house was part of Consolidated Cortez's revamping of the water system and that it was inhabited since at least 1919. Artifacts we found both in the platform area and in nearby refuse dumps supported this conclusion—pieces of tar paper, window glass, linoleum fragments, nails, and milled wood along with light bulb fragments and several electrical insulators.

A depression east of the house was a trash pit. It included a canning jar lid, one toy truck and the wheels from another, a porcelain doll fragment, and a fragment of a phonograph record. At the bottom of the dump, buried several feet underground, was an automobile tire with a patent date of 1917.

A dugout just north of the house had rocks at the corners and extremely compacted soil marking the walls and floor. We found a range of building materials— tar paper, fragments of mortar and brick, nails, and sheet metal. But other artifacts hinted at a residential use. These included a clock spring, a rice bowl fragment, light bulb glass, bone fragments, eggshell, peach pits, saw-cut butchered beef and sheep bones, and chicken bones. There were also fragments of a porcelain doll. It could have been an early residential dugout, perhaps home to the Chinese pump tender of Mabell McElrath's acquaintance, or later the mysterious chicken coop of Estelle Bertrand Shanks's memory.

The pump house residents, before, after, and including the Bertrands, left a constellation of dumps in the sagebrush out and away from their home. These consisted mostly of tin cans and other food waste, narrowly dating from the 1910s to the 1930s and coinciding with the Consolidated Cortez operations.

This information about the Consolidated Cortez pump tenders gave us the means to compare their lives with the previous residents of the pump station and other residents of the Cortez District. The first thing we realized was that, among all this post-1910 debris, even the minimal Chinese artifacts we saw at the rest of the

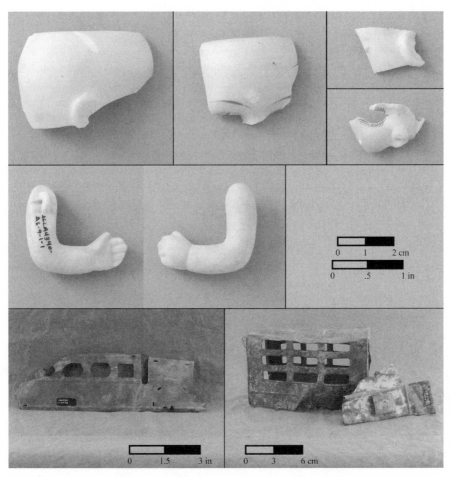

FIGURE 8.3. The domestic side of life at the pump station included dolls and toy trucks. Courtesy of Summit Envirosolutions, Inc.

site were nowhere to be found. The Chinese presence had ended. The numerous food and evaporated milk cans showed an increased reliance on canned food and a preference for dishes such as biscuits, pancakes, or casseroles. Frank and Estelle, or other families with growing kids, might have consumed a lot of milk. There was also a nice variety of other products—salad dressing from New York, mincemeat from Philadelphia, and chili powder from San Antonio, Texas, in addition to canned meat from the United States and South America and canned fish from Norway or France. Garages and mechanics were no closer than medical service, so the pump station families did their own repair work, as indicated by a variety of automobile parts, piston heads, springs, tires, tie rods, and a steering column. Estelle's job, when her father worked on their Chevy, was to stand by and hand him tools as he needed them.

For an archaeologist, it is tempting to make that last connection between an artifact and a specific person, but there is really no way of knowing. Did the broken china doll or tea set belong to Estelle Bertrand? Did her brother play with the toy trucks? Did she? Or did they belong to other children who lived there for a time, moved on with their families, and grew old without happening to record their memories of a little house on the valley floor beneath Mount Tenabo?

The Bertrands moved to the town of Cortez in 1929, as Consolidated Cortez began closing the mine and mill. They lived briefly in a small adobe house and ran the post office and company store. Shortly after, the mill closed completely, ending the town's company-supplied electrical power. Estelle had a very clear memory of the appointed night. Her family, and all the others in town, were ready with their Coleman lanterns, kerosene lamps, and candles. And then the lights went out.

NOTES

1. US Bureau of the Census (1910), Cortez Precinct, Lander County, Nevada.
2. Allen (1995); Bertrand (1995); Shanks (1995).
3. Englebright (2011).
4. Keyser-Cooper (1977).
5. Englebright (2011); *The Northern Miner*, April 15, 1948.
6. Allen (1995).
7. *Reese River Reveille*, August 3, 1907.
8. McElrath (1998).
9. Bertrand (1995); Shanks (1995, 2012–2015).
10. Allen (1995, 80).

Chapter 9

Return to Mill Canyon

Falling silver prices, the stock market crash in October of 1929, and the Great Depression that followed ended the Cortez District's six and a half decades of success. Mining and milling did continue, even making a profit at times, but these scattered, inconsistent endeavors never approached the scale or the returns enjoyed by Simeon Wenban or Consolidated Cortez.

Miners and mill workers felt the effects first. Consolidated Cortez employed sixty workers in 1929,[1] but by the time census takers of the Fifteenth US Census arrived in April 1930 there were none to be counted. The Cortez Precinct had shrunk to twenty-one total inhabitants. There was one miner, a mining engineer, and a man employed as a "mine watchman," presumably overseeing the idle mine and mill. Matthew Bertrand was a storekeeper. A teacher, trucker, and laborer who did odd jobs made up the rest of the population. The Garrison Precinct included three ranching families in the north end of Grass Valley, and no miners.[2]

The Englebrights had moved permanently to Canada, but they stayed connected to Cortez through the Rossi family. They visited during the summer, taking the cross-country train from the Midwest to Beowawe. William Englebright continued on to San Francisco, while Miola, Bill, and his younger sister stayed behind with the Rossis. William Englebright's visit to San Francisco included the yearly gathering of the Bohemian Club, which was, and still is, an exclusive male social club that counted many prominent California and national business figures among its members. Its connection to Cortez goes back to the club's early days. In 1892, Simeon Wenban joined the Bohemian Club and made his lot at Sutter and Mason Streets available for a new building, eventually loaning the club $12,000.[3] The Bohemian Club was a thread linking Wenban and quite possibly the managers of his estate to Consolidated Cortez. Charles Kaeding, another executive and manager for Consolidated Cortez, also belonged to the Bohemians.[4] The club provided just the kind of setting where business relationships could be established and cultivated and deals made. Meanwhile, young Bill Englebright's fondest memories of those summers were the prospecting trips he took with his uncle, Louis Rossi.

The Bertrands' life as a mining family did not markedly change during the Depression. Jobs were scarce, and money scarcer, but Matthew continued to find work wherever he could. They moved to Carlin in 1930, after the mill closed and Cortez emptied out. They returned in 1935, when Matthew began working with John Boitano. He had taken a lease on the old Garrison Mine—also called the No. 1. State records showed the "Boitano Mine" producing 411 tons of ore with a gross value of

FIGURE 9.1. Cortez mine workers, ca. 1920s. Courtesy of Estelle Bertrand Shanks.

$24,997 from 1936 to 1937.[5] Estelle remembered the mine gave them a "pretty good living." "We made it, we got by, and we enjoyed being there." One memorable amusement for the children was riding into No. 1 in a big, mule-drawn ore car and being severely cautioned by the adults to stay on one particular side of the tunnel, as the other side had bottomless drop-offs into the darkness.

In 1937, Estelle and Frank had graduated from the little Cortez school, and the family returned to Carlin so they could attend high school. Estelle was a more successful student than Frank, who, as he said, "went to second year of high school and then I decided not to go anymore."[6] Matthew Bertrand worked on mining projects near Carlin and Imlay. One winter he cut ice from the railroad ponds in Carlin. The blocks of ice were stored and then distributed throughout the summer to refrigerated freight cars on their way across the country. In January 1939 he got a job working in Mill Canyon, first in the Emma E. Mine and then the Roberts Mill, a short way down the canyon from the original 1864 mill. His son Frank joined him later, working in the assay office making $4.50 a day; miners made $5.50.[7] Charlotte and Estelle remained in Carlin for a time, then Charlotte moved into a small cabin about a mile up the canyon from the mill. She worked in the Mill Canyon boardinghouse, cooking and housekeeping. Estelle went to Austin to finish high school, where she lived first with relatives and then family friends.

The Bertrands were by no means the only family in Mill Canyon, nor was theirs the only story we uncovered. During some of the first archaeological work in the Cortez District in 1977, an archaeologist with the Bureau of Land Management recorded a small mining site at the southern edge of Crescent Valley, about a mile outside the mouth of Mill Canyon. The site included a good-size leveled area, a rock wall, and what was presumed to be a dugout. Other archaeologists investigated the site in 1990. They mistook the dugout for a collapsed adit and speculated that it might have been a turquoise mine, since they observed some turquoise nuggets at the site, and there are a number of turquoise deposits in the area.

We revisited the location, and this time it was thoroughly investigated. As we cleared away the dense brush, we gradually exposed an extensive residential site.

The leveled area held the remains of a wood frame building, possibly one of the earliest structures in the Cortez District. There was a very distinct, hard-packed dirt floor about seven inches below the surface and dense with artifacts. These included the many machine-cut nails used to construct the building and an extraordinary array of everyday objects: glass fragments, animal bones, cloth, a shovel, ammunition, belts and belt buckles, buttons, shoes, glass jewelry, pencil lead, a pocket watch, horseshoes, mule shoes, and an 1865 half-dollar.

The rock wall turned out to be a retaining wall supporting the lower portion of a large, split-level structure. A six-foot-high embankment separated the two levels, which combined represented a large boardinghouse or hotel dating prior to the 1890s. There was an extensive accumulation of refuse, including mining tools, work boots, and buttons from work clothing. A china doll fragment, cosmetic jar, and piece of glass cameo jewelry indicated women and children lived at the site. All in all, the evidence showed as many as twenty to thirty miners, and some families, could have lived there at one time.

We also discovered another, smaller residence, with a stone fireplace and, behind it, a small, rock-lined depression. We began excavating the depression thinking it was a privy, or outhouse. It measured about four by seven feet; it was six inches deep; and outlined on the surface by a rectangular arrangement of rocks. We presumed the rocks were the foundation for the privy structure, while the depression resulted from the settling of the privy fill over time. Privies are an important archaeological feature because, aside from their obvious use, they also doubled as trash pits, collecting household debris of all kinds. They often contain medicine and liquor bottles and, needless to say, even important personal items occasionally fell into a privy and were not retrieved.

These expectations changed dramatically when, at a depth of about a foot and a half, we came across a human finger bone. Work stopped for a few hours while we consulted with the Bureau of Land Management on a course of action. The excavation then continued with extreme care, as is the case whenever human remains are encountered. We soon realized we were re-excavating a grave, as the fill was easily distinguished from the surrounding, intact soil. But understanding exactly what we had was not so simple. At a depth of about two and a half feet, we began exposing thin wooden planks lining the excavation. They were parts of a large box, but there were no handles or other fancy casket hardware. It looked like a simple shipping crate, and when we next encountered a small area of horizontal wood we assumed we had reached the bottom of the box. Our "coffin" was apparently empty. This was not entirely unknown at Cortez. We had no evidence of a Chinese association with this grave, but it was customary for the Chinese to exhume the remains of "sojourners" and repatriate them to the home country. Just such exhumations had occurred at Cortez's Chinese cemetery.

We continued working to expose more of the box's wooden floor and then, to our surprise, uncovered what was unmistakably a human skull. Using brushes and wooden tools and acting with the greatest care, we slowly exposed what turned out to be a complete human skeleton. The person's belt was still in place, as were a pair of boots, and remnants of his clothing.

Archaeology is a very interesting thing to do for a living, but like many jobs it does at times become routine. You get up in the morning, at a motel or maybe a campsite near the project area, find some coffee, have breakfast, and make sure you've got lunch and water and all the tools and gear you'll need for the day. You pile into crew cab pickup trucks and take off into the sagebrush toward the site you are excavating. You spend a full day digging or carefully troweling through the dirt, sometimes inside a dugout or some other kind of structure, sometimes out in the open. You screen everything, picking out pieces of broken bottle glass, fragments of tin cans, pieces of ceramic plates, and occasionally something personal or unique— a watch, a doll part, a coin. You take photographs and then sit down and complete what feels like reams of paperwork. It can be like any other job, until it's not.

Finding a human skeleton in the course of an archaeological investigation is actually quite rare. You can spend a career working in the Great Basin without ever encountering a grave, and most archaeologists are just as happy not to have the experience. But it is an occasionally unavoidable part of digging into the past. Because of that, we build into projects procedures to address these unexpected discoveries. There are a number of protocols to be followed, beginning with contacting law enforcement just to eliminate the possibility that you have stumbled upon a modern crime scene. The next step is to establish contact with the next of kin, including appropriate tribal groups if the remains are Native American. The excavation proceeds with special care and with the constant awareness that you are dealing with the remains of a fellow human being.

After continued consultation with the Bureau of Land Management, we removed the bones. We answered the "empty coffin" question when we realized that at some point one end of the box had either been undermined or there had been a cavity below it in the original grave. The wood decayed and gave way, causing the body to shift toward this low point at the foot of the box. This left nothing at the other end and gave us the initial impression that the entire box was empty.

We transported the remains to the anthropology laboratory at the University of Nevada, Reno, designating them as "John Doe" for the time being. It was imperative to find out who the person was, give him his name back, and preferably find a living relative.

FIGURE 9.2. The receipt from the Allen Mercantile in Beowawe and box of .22 cartridges found with the remains of Mr. Brown. Courtesy of Summit Envirosolutions, Inc.

Making the identification from an unmarked grave was going to be a challenge. We found a number of items with the body, but there was no identification. In addition to his boots, we had a pair of leather work gloves, a pocket knife with the hand-carved initials "FNB," a cardboard box of .22 shells, a partially legible receipt from the Allen Mercantile in Beowawe, a coin purse, and a shoelace. Remnants of clothing covered the skull and part of the body.

We removed the skeleton with a minimal amount of cleaning. That would be left to the forensic anthropologists under more controlled laboratory conditions. The skeleton arrived at the lab covered in dirt, with roots and soil still holding many of the bones in place. A genetic test on one of the bones determined the person was of western or central European ancestry, but John Doe's true identity remained unknown.[8] The next surprise came once the thorough cleaning began. As lab workers began removing the dirt and decayed fabric encasing portions of the skull, the cause of death became obvious. The skull had a small hole at the temple on one side, and a larger hole on the opposite side. These were entry and exit wounds caused by a small caliber bullet—the clue that eventually led us to John Doe's real name and story.[9]

We hoped his death would not have gone unreported, so our next stop was the Lander County courthouse where we began working our way through the county's death records. We began with 1930, because the unfired bullets had markings dating to that year. We examined the death certificates one at a time, which was not an overwhelming task in the sparsely populated county. A very promising lead soon turned up. Certificate number 233 documented the death, in 1938, of F. W. Brown. The certificate indicated that there was an eight-day gap between the day of his death and discovery of his body. The cause of death was suicide, by "bullet in the head," and the remains had been buried on a hillside near where they were found.[10]

Armed with a name and date of death, the next step was to review area newspapers to see if the tragedy had made the local news. The *Reno Evening Gazette*, *Elko*

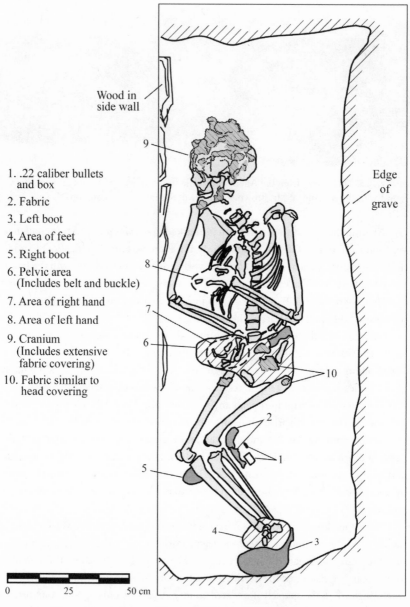

Wood in
side wall

9

1. .22 caliber bullets
   and box
2. Fabric
3. Left boot
4. Area of feet
5. Right boot
6. Pelvic area
   (Includes belt and buckle)
7. Area of right hand
8. Area of left hand
9. Cranium
   (Includes extensive
   fabric covering)
10. Fabric similar to
    head covering

Edge
of
grave

8

7

6

10

2

1

5

4

3

0   25   50 cm

FIGURE 9.3. The unexpected grave. Courtesy of Summit Envirosolutions, Inc.

Daily Free Press, and *Battle Mountain Scout* all reported that the body of a Mr. Brown, age forty-four, had been found in the eastern part of the county. Mr. Brown had been missing from his home for more than a week, according to his wife. Search parties were sent out, but he was not found until someone spotted his car in a small clump of willows. His badly decomposed body was nearby, with a bullet wound to the head. A .22 caliber rifle was at his side, along with a shoelace removed from one of his shoes and looped so that he was able to "pull the trigger with his own hand."[11] The pocket knife, box of shells, and the other items found in the grave were probably in his pockets at this time, and the string may still have been attached to his finger.

The county coroner, a physician, and a constable traveled to the scene, accompanied by three other men constituting a coroner's jury. They held an inquest on the spot, with the cause of death determined to be suicide. Afterward, "it was found necessary to immediately bury the body there [at the place of discovery] because of its decomposition caused by the sun and rain."[12]

The discovery of Mr. Brown's body gave us, at least for one individual in the late 1930s, a blunt answer to the question of what it was like to live in the Cortez District. For him, it was unbearable. The newspapers reported Mr. Brown had been suicidal for some time, and one story concluded he had finally made good on previous threats to kill himself. Friends noted that he had been in ill health for years, a fact our later forensic study confirmed. The newspapers also pointed out that the family was no stranger to tragedy. The gun Mr. Brown used to kill himself was the same one with which, several years earlier, his young son had accidently shot and killed his daughter when they lived in Raleigh, a placering camp in the Shoshone Range. Mr. Brown left behind a widow, two sons, and a second daughter. The initials on the pocket knife—FNB—were his son's.

Mr. Brown was forty-four years old when he took his own life. He was a miner, and the physical analysis of his bones described the effects of a lifetime of hard work. We can only guess at what it may have felt like. The muscle attachments on his bones were robust and well developed, but his upper extremities—his arms and shoulders—showed signs of stress and wear consistent with hammering, shoveling, and pick work. He had low-grade osteoarthritis on most joint surfaces, and his fifth lumbar vertebra exhibited a compression fracture.

He was not taken to an undertaker or given a formal interment in a cemetery. The coroner, constable, and coroner's jury made their determinations, closed the case, and, because of its condition, buried the body where they found it. His coffin was a wooden box, salvaged or put together on the spot. Rocks marked his grave.

On a hot July day in 1938 two cars made their way out from the foot of Mount Tenabo into the southern end of Crescent Valley. They raised a cloud of dust that followed them around to the mouth of Cortez Canyon, where they disappeared from sight. Their duty done, the men returned to the county seat at Austin, filed their report, and went on with their lives. Their grim task was complete, and it must have been a relief to put it behind them. There is no record of what those days were like in the Brown household. For all the things archaeology can tell us about the past, private grief and mourning are not among them.

Six months later, in January 1939, Matthew Bertrand landed a job in Mill Canyon. He lived in the bunkhouse at first, walking up the hill to work at the Emma E. Mine. His wife and son Frank joined him a few months later, but until then he kept in regular touch by letter. These remain valuable chronicles of the workaday life of a Depression-era miner in north-central Nevada. Nor would it be inaccurate to say he wrote essentially the same letters miners had written home in a dozen different languages throughout the history of the Cortez District.

One such letter was dated January 19, 1939, not long after Matthew had been hired:

To my Dear Ones:

Well, I got a job mucking. [Muckers went in after a blast and shoveled loose rock into the ore cars.] It is not bad at all. But, climbing the mountain is fierce. I was about all in when I got on the top this A.M. But think I will make it all O.K. after I get hardened in. [Bertrand had a heart condition.] I had an old Irishman for a side kick today. He is quite jolly. We shoveled dirt and put down ore in twenty feet of track.

There is a night shift working in the mine and they don't change shifts. So, I am lucky to get on the day shift. I am in a room that five are sleeping in.

Another letter is dated Sunday, March 19, 1939:

I went to work here two months ago today. And it seems like a year since I left you folks.

That is all bunk about Murphy saying the ground was safe. Murphy came in two days after the ground caved. [Apparently Murphy had been accused of falsely attesting to the safety of the mine, when in reality he was not there until two days after the cave-in.] Everybody in the mine knew the ground would cave, as there had been about 3,000 tons of ore had been taken out of this stope, you can imagine the size of the hole. Well the machine men put several holes in the [pillar] that was holding the roof and shot the [pillars] of ore out and that let the roof down. As soon as the blasts went off, some of us fellows got in a safe place and watched the roof of the hole fall, which was a lot of rock. And as soon as they took the ore out which was in a reasonable safe place. The boss put in a door and closed the place off. Perhaps some of the fellows that got layed off, was sore and made that report. Someone knocking the co. and Murphy.

Things are not looking too good here lately. The values in the ore has dropped away down. We four men has been putting the stock pile in the bin at the mill, as the ore in the mine is too low grade. If they don't find something better in the near future, I am afraid things won't go so good. (Keep this under your hat. As I don't want it said Matt said it.)

Estelle attended high school in Austin while her father worked in Mill Canyon. She visited often, however. Once while she walked with him, he pointed out the grove of trees where Mr. Brown had shot himself. The Bertrands had not personally known Mr. Brown, but they knew of him, as did everyone in the small community of miners and their families. They knew he had been having a hard time, and of course everyone remembered the tragic story of his daughter's accidental death. Estelle put the ever-present discouragement and loss of hope succinctly in one of her interviews: "It was the Depression. People didn't have *anything*."

The Roberts Mill operated until 1940,[13] but Mill Canyon's potential, having been eclipsed from the beginning by mines on the Nevada Giant side of Mount Tenabo, was never realized. The final shutdown marked the end of significant mining in the Cortez District, a hiatus that would last for decades. The town of Cortez was already well on its way to becoming a classic Western ghost town. A world war was approaching, and in a few years the government would decide that miners' time and energy was better spent producing strategic metals. Silver was a luxury for another day.

NOTES

1. Johnson and McQueen (2016: Mining and Milling Technology, 32).

2. US Bureau of the Census (1930), Nevada, Lander County, Austin Township, Sheet No. 8A. Rural precincts included both towns and surrounding ranches, although the populations could be spread out over many square miles.

3. Bohemian Club (1895, 53); Bohemian Club (1930); *The Morning Call*, May 2, 1892.

4. Bohemian Club (1931).

5. Couch and Carpenter (1943, 61.6).

6. Bertrand (1995).

7. Bertrand (1995).

8. Spencer et al. (2012).

9. Spencer et al. (2012).

10. Nevada State Board of Health, Bureau of Vital Statistics, Death Certificate No. 233, F. W. Brown.

11. *Battle Mountain Scout*, July 28, 1938. Of the three newspapers recounting his death, the *Scout* is geographically closest and provides the most information.

12. *Battle Mountain Scout*, July 28, 1938.

13. McQueen et al. (2015).

Chapter 10

The Silver Never Failed

What was it like to live in the Cortez District?

You might awaken in the morning on the dirt floor of a dugout or in a warm bed at a boardinghouse. Your breakfast might be a handful of rice, or a stack of pancakes, or nothing more than a cup of boiled coffee. Your day might involve hammering drill holes into rock by candlelight or making beds. At day's end, there might be nothing to do but get ready for tomorrow, or warm yourself around a stone oven enveloped in the aroma of freshly baked bread. Sleep might come with pain and bone-deep weariness, or a child's anticipation of the coming day.

There were as many answers to our question as there were people in the Cortez District, and our many years of archaeological and historical study uncovered more than a few of them. We hope we have given you a meaningful history, of both of the district and its people.

And if the Cortez District could say one thing to tell its own story?

The silver never failed.

It made Simeon Wenban a millionaire and kept body and soul together for men like Matthew Bertand and their families. It allowed hundreds of others to make places for themselves, from snug, wood frame homes to half-buried hovels. The unfailing silver gave the Cortez District its extraordinary longevity, and it allowed Wenban to take the uncertainties of finding precious metal and extracting it from solid rock, miles from anywhere, and mold them into a functioning, prosperous enterprise. The silver validated the otherwise silly idea of using mules to haul packs loaded with rocks up one side of a mountain and down the other, or across miles of high desert. It forgave the mistake of freighting the wrong kind of mill piece by piece over the Sierra Nevada to a remote canyon, and then left time for the experiments that finally produced a mill that worked. The silver showed enough of itself to convince Simeon Wenban to spend his entire fortune opening up the Nevada Giant side of Mount Tenabo and to commit even more money to a new, state-of-the-art mill, with its own water system and eventually its own town. The Cortez silver gave Wenban and his family a life equal to any Bonanza King, with enough leftovers in the tailings to support a whole new cyanide leaching industry in the early twentieth century. Then, without missing a beat, Mount Tenabo's silver sustained Consolidated Cortez for almost a decade and, for one year, made Cortez the top producer in the state. In the end, lawsuits, the fickleness of distant markets, and a world war finally turned people away from the precious metal that still lies under the Nevada Giant.

But there was always more ore. After World War II a new generation of miners took interest in the mountain. This time it was gold. Wenban's mill had recovered trace amounts of gold with the silver ore, but the microscopic particles remained largely untouchable. That changed in the 1960s, and in 1969 mines in the Cortez District were among the first in North America to experiment with heap leach technology and precious metal recovery. Their success paved the way for a new era of mining in Nevada that continues today. For the Cortez District, the last mining chapter is yet to be written.

Bibliography

BOOKS, ARTICLES, AND TECHNICAL REPORTS

Angel, Myron, ed. 1881. *History of Nevada, with Illustrations and Biographical Sketches of Its Prominent Men and Pioneers.* Oakland, CA: Thompson and West. Reprinted 1958 by Howell-North, Berkeley, California.

Ataman, Kathryn, Robin Bowers, Erika Johnson, Barbara J. Mackey, Robert McQueen, Jeff Northrup, Meredith Rucks, and Susan Stornetta. 2002. *A Class III Cultural Resources Inventory of Sierra Pacific Resources 345kV Transmission Line.* Report submitted to the Elko, Battle Mountain, and Ely Districts, Bureau of Land Management, Nevada. BLM Reports 1-1973(P), 6-2131(P), 8111-04-99-1309(P). Summit Envirosolutions, Inc. Carson City, Nevada.

Bancroft, Hubert H. 1889. History of the Life of Simeon Wenban: A Character Study. In *Chronicles of the Kings.* Republished in 1892 in *Chronicles of the Builders of the Commonwealth,* Vol. 4, 237–259. San Francisco: The History Company.

Beckwith, E. G. 1855. *Explorations and Surveys for a Rail Road Route from the Mississippi River to the Pacific Ocean, Route Near the 41st Parallel. Exploration and Surveys made under the Direction of the Hon. Jefferson Davis, Secretary of War, by Capt. E.G. Beckwith, 3rd Artillery: F.W. Egloffstein, Topographer for the Route.* Washington, DC: War Dept.

Bennet, Frank P. 1920. Financial Inquiries section: Consolidated Cortez Silver Mines. *United States Investor,* March 20, 1920, 31: 24g–h.

Blake, Kellee. 1996. "'First in the Path of the Firemen': The Fate of the 1890 Population Census, Part 1." *Prologue Magazine,* Spring 1996, 28(1).

Bohemian Club. 1895. *Constitution and By-Laws of the Bohemian Club of San Francisco.* San Francisco: Bohemian Club.

——. 1930. *The Annals of the Bohemian Club: Comprising Text and Pictures Furnished by Its Own Members.* Vol. 4. San Francisco: Bohemian Club.

——. 1931. *Officers, List of Members, House Rules, Grove Rules.* San Francisco: Bohemian Club.

Brott, C. W. 1982. *Moon Lee One: Life in Old Chinatown.* Weaverville, CA: Great Basin Foundation.

Brown, Mrs. Hugh. 1968. *Lady in Boomtown: Miners and Manners on the Nevada Frontier.* Reno: University of Nevada Press.

Bunyak, Dawn. 1998. *Frothers, Bubbles, and Flotation: A Survey of Flotation Milling in the Twentieth-Century Metals Industry.* Denver: National Park Service, Intermountain Support Office.

Bureau of Land Management (BLM). 2008. *Cortez Hills Expansion Project. Final Environmental Impact Statement Vol. II.* Battle Mountain, Nevada: US Department of the Interior, Bureau of Land Management, Battle Mountain Field Office.

Christensen, Jon. 2001. *Nevada's Metal and Mineral Production (1859–1940, inclusive).* Nevada Bureau of Mines and Geology, Geology and Mining Series No. 38, XXXVII (4).

Clewlow, C. William, Jr., and Mary Rusco, eds. 1972. *The Grass Valley Archeological Project: Collected Papers.* Nevada Archeological Survey Research Paper No. 3. University of Nevada, Reno.

Clewlow, C. William, Jr., Helen Fairman Wells, and Richard D. Ambro. 1978. *History and Prehistory at Grass Valley, Nevada*. Institute of Archaeology Monograph 7. University of California, Los Angeles.

Consolidated Cortez Silver Mines Company. 1923. *Administrative Report*. Cortez Mining District records, NC254. Special Collections, University Libraries, University of Nevada, Reno.

———. n.d. Untitled company prospectus. Cortez Mining District Records, NC254. Special Collections, University Libraries, University of Nevada, Reno.

Cortez Mining District. n.d. Claim book. Cortez Mining District records, NC10. Special Collections, University Libraries, University of Nevada, Reno.

Couch, Bertrand F., and Jay A. Carpenter. 1943. *Nevada's Metal and Mineral Production (1859–1940, Inclusive)*. Geology and Mining Series No. 38. University of Nevada Bulletin, November 1, 1943, XXXVII (4). Reno: Nevada State Bureau of Mines and the Mackay School of Mines.

Crum, Steven James. 1994. *The Road on Which We Came*. Salt Lake City: University of Utah Press.

De Quille, Dan (William Wright). 1877. *History of the Big Bonanza*. Hartford, CT: American Publishing Company.

Desert Research Institute. 2013. Monthly Precipitation, Cortez Gold Mine, Nevada. Accessed February 15, 2016. http://www.wrcc.dri.edu/cgi-bin/cliMONtpre.pl?nv1975.

Dickson, H. Leavens. 1939. *Silver Money*. Bloomington, IN: Principia Press.

Dirlilk, Arif. 2001. *Chinese on the American Frontier*. Lanham, MD: Rowman & Littlefield Publishers.

Eissler, Manuel. 1891. *The Metallurgy of Silver*. London: Crosby, Lockwood, and Son. Reprinted 2006 by Elibron Classics.

———. 1901. *The Metallurgy of Silver*. 5th ed. London: Crosby Lockwood and Son; New York: D. Van Nostrand Company.

Elston, Robert G. 1986. Prehistory of the Western Area. In *Great Basin*, edited by W. L. d'Azevedo, pp. 135–148. *Handbook of North American Indians*, vol. 11. W. C. Sturtevant, general ed. Washington, DC: Smithsonian Institution.

Emmons, William H. 1910. *A Reconnaissance of Some Mining Camps in Elko, Lander, and Eureka County, Nevada*. Bulletin 408. Washington, DC: United States Geological Survey.

Emmons, William Harvey, and Frank Cathcart Calkins. 1913. *Geology and Ore Deposits of the Philipsburg Quadrangle Montana*. Department of the Interior United States Geological Survey Professional Paper 78. Washington, DC: Government Printing Office.

Fowler, Don D. 1986. History of Research. In *Great Basin*, edited by W. L. d'Azevedo, pp. 15–31. *Handbook of North American Indians*, vol. 11. W. C. Sturtevant, general ed. Washington, DC: Smithsonian Institution.

Garvin, James L. 1994. Small-Scale Brickmaking in New Hampshire. *The Journal of the Society for Industrial Archaeology* 20, no. 1–2: 19–31.

Genealogy Trails. n.d. Eureka County Marriages 1873–2000. Website accessed February 24, 2015. http://genealogytrails.com/nev/eureka/marr/eurekagrmarrA_D.html.

Gilluly, James, and Harold Masursky. 1965a. *Geology of the Cortez Quadrangle, Nevada*. Bulletin 1175. Washington, DC: US Geological Survey.

———. 1965b. Geologic map and sections of the Cortez Quadrangle, Nevada, accessed February 15, 2016. http://contentdm.library.unr.edu/cdm/ref/collection/hmaps/id/4724/.

Grayson, Donald K. 2011. *The Great Basin*. Berkeley: University of California Press.

Gurcke, K. 1987. *Bricks and Brickmaking: A Handbook for Historical Archaeology*. Moscow: University of Idaho Press.

Hague, James D. 1870. Mining Industry. In *Report of the Geological Exploration of the Fortieth Parallel*, by Clarence King. Professional Papers of the Engineer Department, US Army. No. 18. Washington, DC: Government Printing Office.

Hardesty, Donald L. 1988. *The Archaeology of Mining and Miners: A View from the Silver State.* Special Publications Series, Number 6. Society for Historical Archaeology.

———. 2010. *Mining Archaeology in the American West: A View from the Silver State.* University of Nebraska Press and the Society for Historical Archaeology.

Hardesty, Donald L., and Eugene M. Hattori. 1982. *Archaeological Studies in the Cortez Mining District, 1981.* University of Nevada, Reno and the Desert Research Institute. Submitted to the Bureau of Land Management, Battle Mountain District, Nevada. BLM Report 6–462(P).

———. 1983. *Archaeological Studies in the Cortez Mining District, 1982.* Technical Report No. 12. Bureau of Land Management, Reno, NV.

———. 1984. *Archaeological Studies in the Cortez Mining District, 1983.* Report submitted to the Bureau of Land Management, Battle Mountain District, Nevada.

Harmon, Mella. 2011. Dot Com Serendipity: How Electronic Sources Enrich History. *In Situ: Journal of the Nevada Archaeological Association* 15, no. 4 (fall-winter): 6–7.

Hattori, Eugene M., and Marna Ares Thompson. 1987. Using Dendrochronology for Historical Reconstruction in the Cortez Mining District, North Central Nevada. *Historical Archaeology* 21, no. 1: 60–73.

Hattori, Eugene M., Marna Ares Thompson, and Alvin R. McLane. 1984. *Historic Pinyon Pine Utilization in the Cortez Mining District in Central Nevada: The Use of Dendrochronology in Historical Archaeology and Historical Reconstructions.* Technical Report No. 39. Desert Research Institute, Reno, NV.

Hess, Ronald H., Eugene Faust, Mike Tracy, Becky Weimer, and Terrill Kramer. 1987. *Mining History and Place Names of the Comstock Area, a Field Trip Guidebook, September 19, 1987.* Nevada Bureau of Mines and Geology, University of Nevada, Reno.

Hezzelwood, George W. 1930. *Mining Methods and Costs at the Consolidated Cortez Silver Mine, Cortez, Nevada.* United States Bureau of Mines Information Circular 6327.

High, Lloyd. n.d. Cortez Silver Lead Antimony Mine. Manuscript on file at the Nevada Bureau of Mines and Geology Information Office, University of Nevada, Reno. Nevada Mining District File Index ("Cortez"). Reference 12800021.00.

Hobart, Flora Dean. 1954. Indians of Cortez, Nevada: Mary Hall and Her Family. Manuscript on file, Northeastern Nevada Museum, Elko, Nevada.

Johnson, David. 2008. The Archaeology and Technology of Early-Modern Lime Burning in the Yorkshire Dales: Developing a Clamp Kiln Model. *Industrial Archaeology Review* XX, no. 2: 127–145.

Johnson, Erika, and Robert McQueen, eds. 2016. *Mitigation of the Cortez Hills Expansion Project, Lander and Eureka Counties, Nevada.* Draft Report. BLM Battle Mountain District, Nevada. Report BLM6–2454. BLM Elko District, Nevada. Report BLM1–2532. Summit Envirosolutions Inc., Reno, Nevada.

Jones, Tim. 1995. Brick Clamps. Wall Building Technical Brief. GATE Publications, Federal Republic of Germany.

Keyser-Cooper, Terri. 1977. The Good Old Days. *Battle Mountain Bugle*, October 26, 1977.

Knight, William Henry, ed. 1864. *Bancroft's Hand-book Almanac for the Pacific States, 1864.* San Francisco: H. H. Bancroft and Company.

Krom, S. R. 1885. *Krom's Ore Crushing and Concentrating Machines: Plans for Lixiviation and Concentration works.* New York: S.R. Krom.

———. 1893 *Krom's Ore Crushing and Concentrating Machines: Plans for Lixiviation and Concentration works.* Jersey City, NJ: S.R. Krom.

Lanner, Ronald M. 1981. *The Pinon Pine: A Natural and Cultural History.* Reno: University of Nevada Press.

Ligenfelter, Richard E. 1974. *The Hardrock Miners. A History of the Mining Labor Movement in the American West, 1863–1893.* Berkeley: University of California Press.

Magee, William Flagg. 2010. *The Elephant Hunter.* Authorhouse.

Manufacturer and Builder. 1882. The Worthington Pumping Machinery. *Manufacturer and Builder* 14, no. 2: 28.

McCabe, Allen. 1996. *Cultural Resources Inventory of 1,375 AC for Cortez Gold Mines Area 2 in Eureka and Lander Counties, Nevada.* Submitted to Bureau of Land Management, Battle Mountain District, Nevada. BLM Cultural Resources Report 6–1911(P). Archaeological Research Services, Virginia, City, Nevada.

McCabe, Allen, and Erich Obermayr. 2003. *Cultural Resource Inventory of 570 Ac for Cortez Gold Mine's Pediment Project in Lander County, Nevada.* Submitted to Bureau of Land Management, Battle Mountain District, Nevada. Report BLM6–2369 (P). Summit Envirosolutions, Inc.

McCabe, Allen, and Ronald L. Reno. 1994. *An Archaeological Investigation and Evaluation of 2194 Acres in the Mill Canyon Area Eureka and Lander Counties, Nevada.* Submitted to the Bureau of Land Management Battle Mountain District, Nevada. BLM Report CR6–1507–1(P). Archaeological Research Services, Virginia City, NV.

McElrath, Mabell Paddock. 1998. Scenes from My Childhood in the Mining Camp of Cortez, Nevada. *Northeastern Nevada Historical Society Quarterly* 98, no. 2: 44–52.

McQueen, Robert, Stephanie Livingston, and Kathryn Ataman. 2008. *On the Margins of Upper Cortez: A Class III Cultural Resources Inventory in Grass Valley and Cortez Canyon, Lander County, Nevada.* Submitted to Bureau of Land Management, Battle Mountain District, Nevada. Cultural Resources Report 6–2494(P). Summit Envirosolutions, Inc., Carson City, Nevada.

McQueen, Robert, Jennifer Sigler, and C. Shaun Richey. 2015. *Mill Canyon Corridor Project: A Class III Cultural Resources Inventory of 2,576 Acres in Mill Canyon, Cortez Mining District, Lander and Eureka Counties, Nevada.* Submitted to Bureau of Land Management Elko District, Tuscarora Field Office, Nevada, Report BLM1–3060(P), and Bureau of Land Management Battle Mountain District, Mount Lewis Field Office, Report BLM6–3038–2(P). Summit Envirosolutions, Inc., Carson City, Nevada.

Megraw, Herbert A. 1918. *The Flotation Process.* New York: McGraw-Hill.

Mifflin, M. D., and M. M. Wheat. 1979. *Pluvial Lakes and Estimated Pluvial Climates of Nevada.* Nevada Bureau of Mines and Geology Bulletin No. 94, 57.

Murbarger, Nell. 1959. Ghost Town: Cortez, Nevada, Population 1. *Desert Magazine* 22, no. 5 (May): 9–13.

———. 1963. The Last Remaining Light. *True West*, January–February, 32–34, 46.

Natural Resources Conservation Service, US Department of Agriculture. n.d. *Nevada Annual Precipitation.* Accessed February 15, 2016, http://www.wrcc.dri.edu/pcpn/prism/nv.jpg.

Nevada State Legislature. 1875. *Biennial Report of the State Mineralogist of the State of Nevada: For the Years 1873 and 1874.* John J. Hill. Carson City, NV: State Printer.

Parker, Tom C. n.d. *Report on the Cortez Mine.* Cortez Mining District Records, NC10. Special Collections, University Libraries, University of Nevada, Reno.

Patterson, Edna. n.d. *Mary Hall, the Western Shoshone Basketmaker, and Her Daughters.* Manuscript on file at Northeastern Nevada Museum. Elko, Nevada.

Perry, Frank, Robert Piwarzyk, Michael Luther, Alverda Orlando, Allen Molho, and Sierra
Perry. 2007. *Lime Kiln Legacies: The History of the Lime Kiln Industry in Santa Cruz County*.
Santa Cruz, CA: Museum of Art and History.

Praetzellis, Mary, and Adrian Praetzellis, eds. 2009. *South of Market: Historical Archaeology
of 3 San Francisco Neighborhoods, the San Francisco–Oakland Bay Bridge West Approach
Project*. Prepared for California Department of Transportation, District 4, Oakland.
Anthropological Studies Center, Sonoma State University, Rohnert Park, California.

Procter, Ben. 1998. *William Randolph Hearst: The Early Years, 1863–1910*. Oxford: Oxford
University Press.

Raymond, Rossiter. 1869. *The Mines of the West: A Report to the Secretary of the Treasury*. New
York: J.B. Ford and Company.

Rehels, Marith. 1999. *Extent of Pleistocene Lakes in the Western Great Basin*. US Department
of the Interior, US Geological Survey. Miscellaneous Field Studies Map MF-2323. Accessed
March 7, 2016. http://pubs.usgs.gov/mf/1999/mf-2323/mf2323.pdf.

Reno, Ronald L. 1994. *The Charcoal Industry in the Roberts Mountains, Eureka County, Nevada:
Final Report for the Mitigation of Adverse Effects to Significant Cultural Properties from Atlas
Precious Metals' Gold Bar II Mine Project*. Submitted to Bureau of Land Management, Battle
Mountain District, Nevada. Cultural Resources Report 6–1193. Archaeological Research
Services, Inc., Virginia City, Nevada.

———. 1996. *Fuel for the Frontier: Industrial Archaeology of Charcoal Production in the
Eureka Mining District, Nevada 1869–1891*. Unpublished PhD dissertation. Department of
Anthropology, University of Nevada, Reno.

Ring, Orvis. 1909. *State of Nevada Biennial Report of the Superintendent of Public Instruction
1907–1908*. Carson City, NV: State Printing Office.

Roberts, Ralph J., Kathleen M. Montgomery, and Robert E. Lehner. 1967. *Geology and Mineral
Resources of Eureka County, Nevada*. Nevada Bureau of Mines Bull. 64. Mackay School of
Mines. University of Nevada, Reno.

Rucks, Meredith. 2000. *A Report on Ethnographic Study Conducted to Assist the Bureau of Land
Management in the Evaluation of Traditional Cultural Properties in the Mt. Tenabo Area of
Lander and Eureka Counties, Nevada*. Submitted to the Bureau of Land Management, Elko
District. Report BLM6–2174–1. Summit Envirosolutions. Carson City, Nevada.

———. 2004. *An Ethnographic Study Completed for the Cortez Gold Mines Pediment Project*. Submitted to ENSR, Fort Collins, Colorado. Summit Envirosolutions, Inc., Carson City, Nevada.

———. 2008. Appendix A in *On the Margins of Upper Cortez: A Class III Cultural Resources
Inventory in Grass Valley and Cortez Canyon, Lander County, Nevada*. Submitted to Bureau
of Land Management, Elko District Office, Nevada. BLM Cultural Resources Report
6–2494(P). Summit Envirosolutions, Inc., Carson City, Nevada.

Schlereth, Thomas. 1991. *Victorian America: Transformations in Everyday Life, 1876–1915*. New
York: HarperCollins.

Simpson, J.H. 1876. *Report of Explorations Across the Great Basin of the Territory of Utah for a
Direct Wagon-Route from Camp Floyd to Genoa, in Carson Valley, in 1859, by Captain J. H.
Simpson*. Washington, DC: Government Printing Office.

Spence, Clark C. 2000. *British Investments and the American Mining Frontier, 1860–1901*.
London: Taylor and Francis.

Spencer, Diana, Robert McQueen, Penny Siig, and G. Richard Scott. 2012. *Back from the Dead:
An Osteobiography of a Depression-Era Miner from Central Nevada*. Poster presented at the
81st annual meeting of the American Association of Physical Anthropologists, Portland,
Oregon.

Stetefeldt, Carl August. 1888. *The Lixiviation of Silver-ores with Hyposulphite Solutions: With Special Reference to the Russell Process.* New Haven, CT: Press of Tuttle, Morehouse & Taylor.

Steward, Julian. 1938. *Basin-Plateau Aboriginal Groups.* Bureau of American Ethnology Bulletin 120. Washington, DC: Smithsonian Institution.

————. 1943. *Some Western Shoshone Myths.* Anthropological Papers 31. Bureau of American Ethnology Bulletin 136: 249–299. Washington. DC.

Tingley, Joseph V. 1992. *The Mining Districts of Nevada.* Issue 47 of Report (Nevada Bureau of Mines and Geology), Mackay School of Mines, University of Nevada, Reno.

United States Bureau of the Mint. 1883. *Report of the Director of the Mint Upon the Statistics of the Production of the Precious Metals in the United States.* Washington, DC: Government Printing Office.

United States Department of the Interior, Geological Survey. 1938 (reprinted 1970). *Nevada Cortez Quadrangle 15-Minute Series.* Washington, DC.

Vanderburg, W. O. 1938. *Reconnaissance of Mining Districts in Eureka County, Nevada.* Information Circular 7022. Washington, DC: US Bureau of Mines.

Weed, W. 1922. *The Mines Handbook, vol. 15.* Tuckahoe, NY: The Mines Handbook Company.

Wegars, Priscilla. 1993. *Hidden Heritage: Historical Archaeology of the Overseas Chinese.* Amityville, NY: Baywood Publishing Company.

Woolley, Dale E. 1999. *The Waltis and Other Early Settlers of Eureka and Lander County, Nevada.* Published by the author.

Wyckoff, William, and Larry M. Dilsaver, eds. 1995. *The Mountainous West.* Lincoln: University of Nebraska Press.

Wyman. Mark. 1979. *Hard Rock Epic: Western Miners and the Industrial Revolution, 1860–1910.* Berkeley: University of California Press.

Yancey, John. 1998. The Last Family in Cortez. *Northeastern Nevada Historical Society Quarterly* 98, no. 2: 44–52.

Young, James A., and Jerry D. Budy. 1979. Nevada's Pinyon-Juniper Woodlands. *Journal of Forest History* 23, no. 3: 113–121.

HISTORICAL RECORDS AND DOCUMENTS

Cortez Company Store Ledger. February–April 1888. Special Collections, University Libraries, University of Nevada, Reno. NC253 Box 2.

————. 1913–1914. Special Collections, University Libraries, University of Nevada, Reno.

————. November 1889–January 1890. Special Collections, University Libraries, University of Nevada, Reno. NC253 Box 2.

————. 1867–1886. Special Collections, University Libraries, University of Nevada, Reno. NC253 Box 1.

Tenabo Mill and Mining Company Payroll Ledger. June 1896–March 1908. Special Collections, University Libraries, University of Nevada, Reno. NC166 Box 3.

US AND STATE CENSUSES

1863 Nevada State Census, Cortez, Lander County.

1870 US Census, Cortez District, Lander County, Nevada.

1880 US Census, Grass Valley, Lander County, Nevada.

1880 US Census, Beowawe, Eureka County, Nevada.

1900 US Census, Cortez Precinct, Lander County, Nevada.

1900 US Census, Garrison Mine Precinct, Eureka County, Nevada.
1900 US Census, San Francisco, California, Enumeration District 227, Sheet 12.
1910 US Census, Cortez Precinct, Lander County, Nevada.
1910 US Census, Garrison Mine Precinct, Eureka County, Nevada.
1920 US Census, Beowawe Precinct, Eureka County, Nevada.
1920 US Census, Cortez Precinct, Lander County, Nevada.
1930 US Census, Austin Township, Lander County, Nevada.
1930 US Census, Garrison Precinct, Eureka County, Nevada.

NEWSPAPERS AND PERIODICALS

The Anglo American Times
Battle Mountain Scout
Daily Alta California
Daily Reese River Reveille
Daily Territorial Enterprise
Elko Free Press
Mining and Scientific Press
Nevada State Journal
The Northern Miner
Reese River Reveille
Sacramento Daily Record
Sacramento Daily Record-Union
Sacramento Daily Union
San Francisco Call/The Morning Call
White Pine News

ORAL HISTORIES

Allen, Clarence Garfield "Doc." 1995. Lander County Oral History Project. University of
    Nevada Oral History Project.
Bertrand, Frank. 1995. Lander County Oral History Project. University of Nevada Oral History
    Project.
Englebright, William Rossi. 2011. Interviewed by Erich Obermayr. Mill Brae, CA, February 9,
    2011.
Shanks, Estelle Bertrand. 1995. Lander County Oral History Project. University of Nevada Oral
    History Project.
———. 2012–2015. Interviewed on several occasions by Erich Obermayr, Austin, Nevada,
    March 2012 to March 2015.

Index

Page numbers in *italics* refer to figures.